Systematic Biology

Handbook of Animal Types

Kathleen Cratchley

Longman

LONGMAN GROUP LIMITED
London
Associated companies, branches and representatives throughout the world

First published 1980
ISBN 0 582 35324 6

Designed by DP Press Ltd
Set in IBM Baskerville & Theme
by Type Practitioners Ltd.

Printed and bound in Great Britain by
William Clowes (Beccles) Limited,
Beccles and London

Contents

Preface

Recent years have seen a radically changed approach to the study of biology in schools. This has been due to the enormous advances that have taken place during the last decades, especially in the fields of genetics, cytology, microbiology, biochemistry and molecular biology.

The broad sweep of modern biology, covering such diverse topics and involving many other disciplines, can no longer be embraced by the traditional systematic approach, and school textbooks are now organised along functional lines. Such an approach succeeds not only in revealing the dynamic and exciting nature of the subject but also presents modern biology as an integrated subject, emphasising the unity underlying the diversity of life. Classification, however, has become a neglected topic in modern A-level biology textbooks.

The purpose of this handbook is to provide a concise survey of the animal kingdom, following traditional systematic lines. It details the criteria used in the classification of animals and will, I hope, serve as a useful reference book for advanced level students of biology, especially those who find themselves at a loss without sufficient factual details at hand.

Although the drawings in the book are designed to illustrate structure and are consequently diagrammatic and stylized, they can be used to advantage in the interpretation of slides and specimens in the practical class.

1. The Principles of Classification

Introduction

Before any logical study of living organisms can be made, it is obviously necessary to devise some means of arranging this great diversity of different forms into groups of individuals sharing common characteristics, i.e. it is necessary to classify them. There is no inherently 'correct' method of classifying living things since all systems of classification are purely arbitrary and are adopted by man for his convenience. However, during the eighteenth century a scheme of classification and nomenclature gradually emerged which has become universally accepted. It is based on the principle of homology, all organisms sharing basic structural similarities being grouped together. Thus all animals possessing the basic similarities of skull, vertebral column, tail and two pairs of limbs are placed together in the group *Vertebrata,* regardless of any other differences between them. Each major group is then sub-divided into smaller groups composed of organisms with additional features in common. This method of classification has become known as the 'Natural System' since it reflects the relationships of organisms, expressing the natural affinities between them. When the system originated, evolution was not an accepted concept, and the recognition of attributes indicating natural affinities between different species was not easy to explain other than being part of the plan of creation. But once general belief in evolution became established it was realised that the more characteristics different organisms share with each other, the more closely related they are by descent; they share a relatively recent common ancestor. In other words the natural system of classification is phylogenetic, and ideally the hierarchy of categories provides a framework reflecting the evolutionary history of living organisms. However a scheme of classification which gives a true picture of the relationships between all the groups is an ideal which cannot possibly be realised; it is bound to be oversimplified and presumptuous. The classificatory system accepted today may be said, at best, to give an indication of the broad trends of evolution. There is no absolute system of classification; differences of opinion exist between biologists and much speculation is involved. The variability of the living world is essentially dynamic and cannot be contained adequately in any static system of categories. The scheme of classificcation must allow for revisions which are continually necessary as evolution progresses and as more is learnt about organisms and about fossil history. For example, advanced physiological and biochemical techniques can sometimes be successful in determining affinities where structural features are inconclusive.

Sometimes for special purposes it is found convenient to classify particular organisms *artificially* as opposed to naturally. For example, a scientist studying methods of locomotion may group together all flying animals, all swimming animals, all burrowing animals etc. Similarly when it is necessary to identify specimens in the field it is usual to make use of keys. These are constructed purely for practical purposes, and are concerned with the identification of specimens from a restricted range of organisms, often the fauna of a particular area. The alternative categories employed to distinguish between organisms may separate them into artificial groupings and may make use of analogous structures. Thus the first step in the identification of the different insects of a particular habitat might well be distinction according to size, which is very necessary for the identification of individual specimens but is irrelevant with regard to their natural classification.

Principles of classification

Animal classification (See pages 18–19)

Classification involves the arrangement of organisms into a hierarchy of groups on the basis of similarities and differences. Since differences between organisms cannot be measured, there can be no clear-cut rules to establish the limits of the different categories. The creation of the groups must therefore be a matter of human judgement, and they cannot be defined precisely. The Swedish botanist Linnaeus devised a 'natural system' of classification in the 1850s. Despite its underlying arbitrariness this still is applied effectively today. Internationally agreed rules ensure that taxonomic procedure remains standardized throughout the world.

The animal kingdom is divided up firstly into major groups, **phyla**, of animals showing radically different plans of organisation, e.g. Coelenterata, Platyhelminthes, Annelida, Arthropoda, Mollusca, Echinodermata and Chordata. Although all the members of the same phylum possess a fundamentally similar body plan, considerable diversity exists amongst them and each phylum is readily sub-divided into different **classes**. E.g. the phylum Arthropoda is divided into the classes Crustaceae, Chilopoda, Diplopoda, Arachnida and Insecta which show significant variations. Similarly in the basic plan each class is further divided into **orders** of animals which have additional features in common, e.g. the class Arachnida includes the orders Scorpions, Spiders, King-crabs and Mites. The various members of an order share the full complement of characteristics attributable to the phylum and to the class to which they belong as well as those common to their particular order. This illustrates the fact that classification is more than just a system for bringing order out of chaos; any information regarding the systemstic position of an organism at once implies many of its characteristics.

Orders are sub-divided into **families**, and at this level distinctions are less obvious, e.g. the different families of spiders are all basically very similar in appearance.

Each family is divided into several **genera** and finally each genus is divided into a number of **species**.

Thus, to give an example, a complete classification of *Pieris brassica* (the Large Cabbage White Butterly) is shown on the left.

Kingdom Animalia
Phylum Arthropoda
Class Insecta
Order Lepidoptera
Family Pierida
Genus Pieris
Species: *Pieris brassica*

It is from these last two categories that an organism derives its scientific name, according to an internationally accepted system of **binominal nomenclature** (first used by Linnaeus). The name is always in Latin and is in two parts, indicating firstly the genus to which the organism belongs and then the particular species. For example: both lions and tigers belong to the same genus, *Panthera*; lions are known as *Panthera leo* while tigers are known as *Panthera tigris*, and these names apply throughout the world.

Sometimes distinct **varieties** are recognisable within a species, e.g. the different breeds of dog. In certain cases further categories may be necessary, such as **sub-phylum, sub-class**, etc.

Only at the species level are the groups delimited naturally, since a species can, in most cases, be said to be a potentially interbreeding group of organisms which are reproductively isolated from all other species. In other words the members of a species are able to breed amongst themselves but are incapable of breeding successfully (i.e. producing fertile offspring) with the members of any other species.

Since evolution is currently active, the classical concept of the fixed species cannot hold true. Also many circumstances exist which make a precise definition of species impossible, e.g. problems occur when organisms habitually reproduce either asexually, parthenogenetically or by self-fertilization, and when species previously recognised as distinct are found to be connected by intermediate forms over wide areas.

2. The Kingdom Protista

In the past all organisms have been classified into either the Plant or the Animal Kingdom. If difficulties were encountered, an organism was assigned to the kingdom which seemed the more appropriate. This is why Bacteria, Fungi, Lichens and Algae are usually included in the Plant Kingdom (where they are grouped together as the Phylum/Division Thallophyta, although it is realised that they have very little in common apart from the fact that all are known to be primitive). Similarly the Protozoa are usually incorporated in the Animal Kingdom, even though some of the Flagellata exhibit many plant-like features.

It is an undisputed fact that all these groups have had a very long evolutionary history and were probably in existence long before definite plants and animals had evolved. Therefore it is now being increasingly argued that there should be at least one more kingdom, distinct from both plants and animals, and the name Protista has been suggested. The main significance behind the establishment of such a third kingdom is to bring about a shift of emphasis, raising the status of these primitive groups.

With the advent of the electron microscope and the sciences of microbiology and biochemistry, new fields of biology have been opened up where the need to classify organisms as either plants or animals is just not relevant.

All schemes of classification are merely for convenience and obviously it will still sometimes be necessary to divide biology into botany and zoology. When this is so, the practice of confining the study of Bacteria, Fungi and the various Algae to botany, and that of Protozoa to zoology can still continue, even though all these groups can now be considered members of the Kingdom Protista.

The Kingdom Protista would comprise the following Divisions/Phyla:

- Viruses*
- Bacteria*
- Blue-green algae (Cyanophyta)
- Slime moulds (Myxomycophyta)
- Fungi
- Lichens
- Green algae (Chlorophyta)
- Golden algae (Chrysophyta)
- Brown algae (Phaeophyta)
- Red algae (Rhodophyta)
- Diatoms (Bacillariophyta)
- Dinoflagellates (Pyrrophyta)
- Euglenophyta* (=class Flagellata of the Protozoa)
- Protozoa*

It should be pointed out that this scheme of classification has by no means been generally accepted and there is still considerable controversy among biologists concerning many aspects of classification, particularly the classification of these organisms of simple biological organisation.

* Included in this book.

3. Viruses

Viruses are sub-microscopic obligate intracellular parasites, variable in size and shape, but in all cases consisting merely of nucleic acid enclosed in a proteinaceous envelope. The fact that many viruses can be obtained in crystalline form indicates the relative simplicity of their structure. All viruses are exceedingly small, ranging in size from about 20nm - 300nm, and they are able to penetrate filters that prevent the passage of bacteria. Some are rod shaped, some spherical, some brick shaped, while some (bacteriophages) are more elaborate, the protein coat being prolonged into a complex hollow tail.

Viruses are classified into three groups according to their hosts, i.e. animal viruses, plant viruses, and bacterial viruses (bacteriophages). They are very highly specific, not only to host, but to a particular tissue within a specific host. Animal viruses contain nucleic acid in the form of DNA or RNA. They cause many diseases, including measles, smallpox, chicken pox, warts, rabies, poliomyelitis, influenza, yellow fever, myxomatosis and foot and mouth disease. Plant viruses contain nucleic acid only in the form of RNA; they cause many mosaic diseases. Bacteriophages contain only DNA.

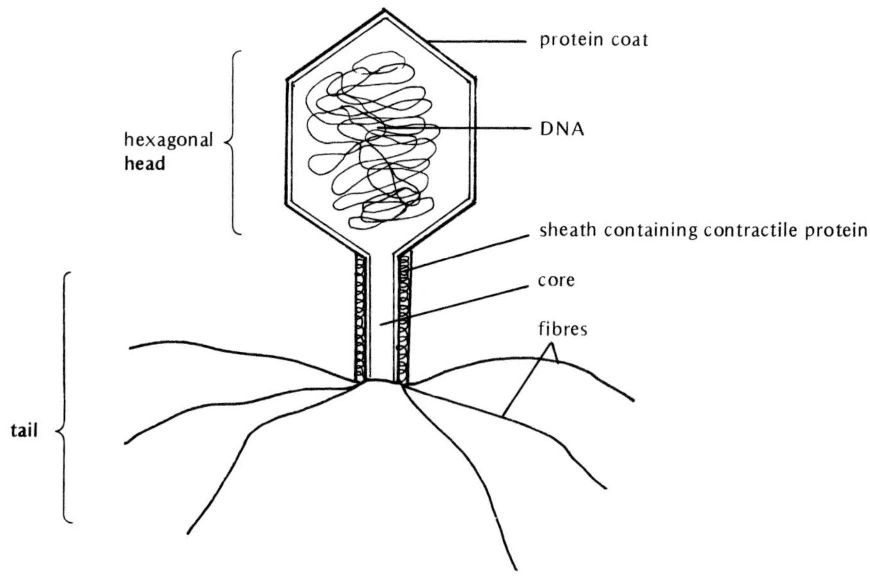

Diagram to show the structure of a bacteriophage (Very greatly enlarged)

Electronmicrographs have shown that when infecting a host cell each bacteriophage becomes attached to the surface of the bacterial cell by its tail fibres, probably due to a serological reaction. Then the sheath contracts driving the core through the bacterial cell wall. The DNA of the phage head then enters the bacterium via the core, leaving the empty protein coat outside.

Until they are in a position to infect a suitable host cell, viruses are biologically inert, showing no signs of any metabolic activity. But once they have gained entry into a cell they are able, by means of their nucleic acid, to take over the protein synthesis mechanism of the cell, whose activities thus become redirected to bringing about the manufacture of new viruses. The mitochondrial activity of the host cell supplies the energy needed, since the virus itself possesses none of the protoplasmic attributes normally associated with cells, other than nucleic acid. New viruses are thus manufactured by the activities of the host cell, and not by fission of a pre-existing virus. The host cell eventually undergoes lysis, and releases the viruses, which can then infect further cells. The production of 'interferon' by the infected cells can sometimes succeed in inhibiting the multiplication process. Viruses can be cultured only in living cells/tissues, not on nutrient jelly, and they are thus not as easily investigated as bacteria.

Once the host's natural defences (antibody production, etc.) have been overcome and infection has taken place, viral diseases cannot be treated by antibiotic drugs, because viruses are intracellular parasites. But resistance to infection by some viral diseases can be achieved by vaccination. If interferon could be produced commercially it would provide an excellent weapon to combat viral diseases, since it is non-specific (i.e. effective against all viruses).

Outside the host, some viruses may be destroyed by heat, ultraviolet light, and various disinfectants. But many are very difficult to eliminate.

Since viruses are completely dependent upon living cells, it is unlikely that they represent a very primitive stage of evolution intermediate between living organisms and non-living materials. They have probably been derived originally from living cells.

4. Bacteria

Bacteria are probably the most numerous and widely distributed of all living organisms. They are microscopic or ultra-microscopic single-celled organisms, representing the smallest forms of life organised on a cellular basis. Besides being exceptionally small (approximately 0.5μm - 4μm in length) the cells are atypical in several respects. They are of a type known as *procaryotic*. In such cells the genetic material (DNA) is not incorporated in chromosomes, and is not segregated within a nuclear membrane. Thus there is no distinct nucleus. Also no spindle forms at cell division. Membrane-bounded organelles (e.g. mitochondria, plastids, endoplasmic reticulum and Golgi apparatus, etc.) are absent, though complex involutions of the plasma membrane may be present. No large vacuoles occur. Procaryotic cells possess cell walls which are unusual in composition – those of bacteria consist of a unique form of polysaccharide and protein.

Although varying greatly in physiology, bacteria occur in only three basic shapes: rod shaped (these are known as bacilli – they are often flagellated), spherical (these are known as cocci), and spiral (these are known as spirilla, and may be flagellated). Their shape is due to the presence of a more or less rigid cell wall. Many bacteria are also surrounded by a gelatinous capsule. Bacteria are motile only if they possess one or more flagella. The flagella are much simpler in structure than those found in eucaryotic cells. They contain only one fibril, whereas those of eucaryotic cells contain eleven, arranged in a characteristic 9 + 2 pattern. Some spirally twisted bacteria (spirochaetes) have a delicate elastic cell wall, and these are able to swim by flexions of the cell brought about by the contractions of two groups of flagella-like fibrils anchored at each end of the bacterium and wound spirally round the protoplast within the cell wall.

Bacteria exhibit a wide range of growth habit. Many remain as individual separate cells; others form filaments or clusters of individuals; in a few, cell organisation breaks down to give rise to a filamentous coenocytic mycelium.

Due to the fact that they have cell walls but no chlorophyll, most bacteria are either parasitic, saprophytic or symbiotic. Between them they possess such a wide range of enzymes that there is no natural organic material that cannot be digested by them. However, some bacteria are autotrophic. But, unlike green plants, most of these exploit energy-yielding inorganic chemical reactions independent of light, i.e. they are chemo-autotrophic. Thus nitrifying bacteria oxidise ammonia to nitrate + energy; hydrogen/methane bacteria oxidise hydrogen to methane, water + energy; iron bacteria oxidise ferrous compounds to ferric hydroxide + energy; and colourless sulphur bacteria oxidise hydrogen sulphide to sulphur + energy. A few bacteria do possess a simple type of chlorophyll (bacteriochlorophyll) and are able to photosynthesise e.g. the green and purple sulphur bacteria. However, they are not able to use water as a source of hydrogen; they must obtain it from hydrogen sulphide. Therefore sulphur, not oxygen, is produced as a by-product. Many of these bacteria are anaerobic.

Bacteria occur in enormous numbers in suitable conditions. Multiplication is by simple fission, and since this can take place as often as once every twenty minutes, an astronomical increase in numbers can occur in favourable surroundings. Sexual reproduction is very rare; a form of conjugation has been observed in a few forms, e.g. *Escherichia coli*, a normally harmless species of bacterium found in the human large intestines.

Unfavourable conditions result in spore production by some species of bacteria; the genetic material etc. becomes enclosed by a wall to form a resting body (which may be greater in diameter than the cell in which it is produced). The spores are easily dispersed and are extremely resistant. However, very few species of pathogenic bacteria can form spores. This is why the pasteurisation of milk kills the *harmful* bacteria in it.

Unlike viruses, pathogenic bacteria are intercellular parasites. Disease symptoms are often caused by the toxins excreted by the bacteria. Nowadays cases of food poisoning are usually due to either the presence of extremely poisonous toxins in food contaminated by *Clostridium botulinum* – the toxins are not destroyed by cooking the food, although the bacteria are – or to the presence of living bacteria in badly prepared food, especially Salmonella in frozen poultry that has been roasted without adequate thawing beforehand. A few species of bacteria are thermophilic, i.e. they have a higher optimum growth-temperature than is usual (above 45°C). Such bacteria occur in hot springs, and in damp hayricks where they can produce spontaneous combustion.

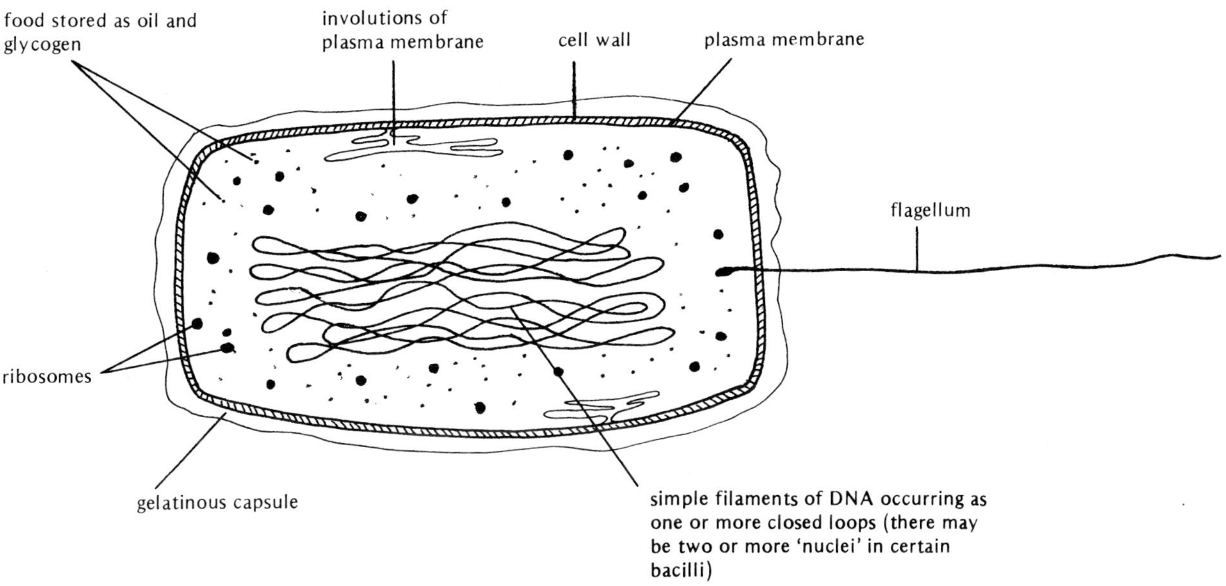

A generalised bacterial cell (Very greatly enlarged)

The activities of bacteria are of great importance

1. Decay and putrefaction
The decomposition of plant and animal remains, etc. is brought about mainly by saprophytic bacteria in the soil and in water. Their importance in this respect cannot be over-emphasised: without these bacteria all the essential materials of life would soon be locked up in the remains of the dead. The speed of decay depends on the presence of sufficient bases, otherwise the conditions become toxic and bacterial activity slows down or ceases. The **disposal of sewage** is dependent on similar bacterial activity.

2. Enriching the soil with nitrogen
N-fixing bacteria living freely in the soil (Clostridium and Azotobacter) and symbiotically in the root nodules of leguminous plants (Rhizobium) are the only living organisms, other than certain blue-green algae, which can utilize atomspheric nitrogen for the synthesis of proteins. When these organisms die, nitrates are added to the soil by the usual decomposition processes.

3. In science and medicine

i. Bacteria are easily cultured and are used by research biochemists working on many fundamental biological problems.
ii. A few bacteria are a source of antibiotics (e.g. Streptomycin).
iii. Bacterial cultures also provide a source of enzymes, etc. Enzymes can be extracted from cultures for use in the manufacture of biological detergents, etc. It is possible that bacterial cultures may soon provide a new source of food.

4. Digestion in herbivores

It is bacteria (and protozoa) in the caecum or rumen of herbivores, and in the intestine of snails, termites, etc. which produce the cellulases necessary for the breakdown of the large amounts of cellulose in the diets of these animals.

5. Cheese and vinegar production

Bacteria are responsible for the souring of milk in cheese production, and for the conversion of alcohol into vinegar.

6. Animal diseases

Bacteria do not cause many plant diseases, but they cause very many animal diseases, e.g. the following human diseases: diphtheria, pneumonia, rheumatic fever, scarlet fever, bubonic plague, whooping cough, cholera, tuberculosis, typhoid, venereal diseases, blood poisoning, boils, gangrene, leprosy.

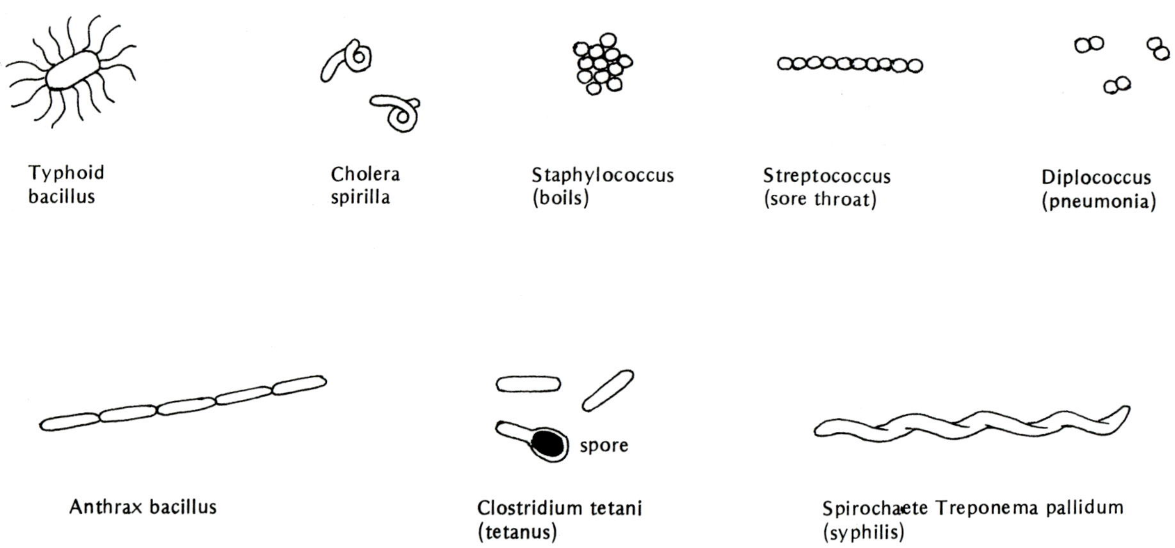

Examples of pathogenic bacteria (Not to scale)

5. Phylum Protozoa

Introduction

The Protozoa include a great diversity of minute single-celled organisms, and some colonial forms. They exhibit more variation in form and physiology than any other single phylum. Protozoa are very widely distributed (ubiquitous) but are active only in aqueous surroundings. Each individual is termed 'acellular' as opposed to unicellular, since there is doubt whether the individual protozoon is homologous with a single cell of a metazoan.

General characteristics

Each individual is comparable with a single cell, usually 2μm - 10μm long.

Well-defined organelles, delimited by unit membranes, are present, including at least one nucleus, i.e. the cell is eucaryotic.

There is no cell wall external to the protoplast, and chlorophyll is absent except in some flagellates.

Nutrition is heterotrophic and typically holozoic, i.e. organic materials are required from the environment, and usually solid food is ingested.

Class Rhizopoda	**Class Mastigophora (Flagellata)**	**Class Ciliata (Ciliophora)**	**Class Sporozoa**
Mostly free-living, few parasitic	Free living and parasitic forms	All free living	All parasitic
Body not very highly differentiated. Some attain a relatively large size, e.g. Amoeba proteus	Body not very highly differentiated. All small in size.	Body very highly differentiated. Some attain a large size.	Structurally simple Size variable
Single nucleus	Single nucleus	Two nuclei: meganucleus and micronucleus	Single nucleus
Move by pseudopodia	Swim by one or more flagella	Move by cilia	No special mode of locomotion
No pellicle — shape irregular unless an external/internal skeleton is present	Definite shape — pellicle present	Definite shape — pellicle present	Definite shape — pellicle present
Food captured and ingested by means of pseudopodia	Nutrition very variable	Food captured by cilia	Soluble food absorbed via body surface
Asexual reproduction by binary fission. Sexual reproduction, involving gametes, in a few species	Asexual reproduction only — by longitudinal binary fission.	Asexual reproduction by transverse binary fission. Sexual reproduction involving conjugation	Asexual reproduction by multiple fission resulting in great number of spores. Sexual reproduction also occurs

E.g. the following genera:

Amoeba **Foraminifera**	**Euglena** **Peranema** **Trypanosoma**	**Paramecium** **Vorticella**	**Monocystis** **Plasmodium**

Phylum Protozoa
Class Rhizopoda
Genus *Amoeba*

CLASS RHIZOPODA

e.g. *Amoeba sp.*

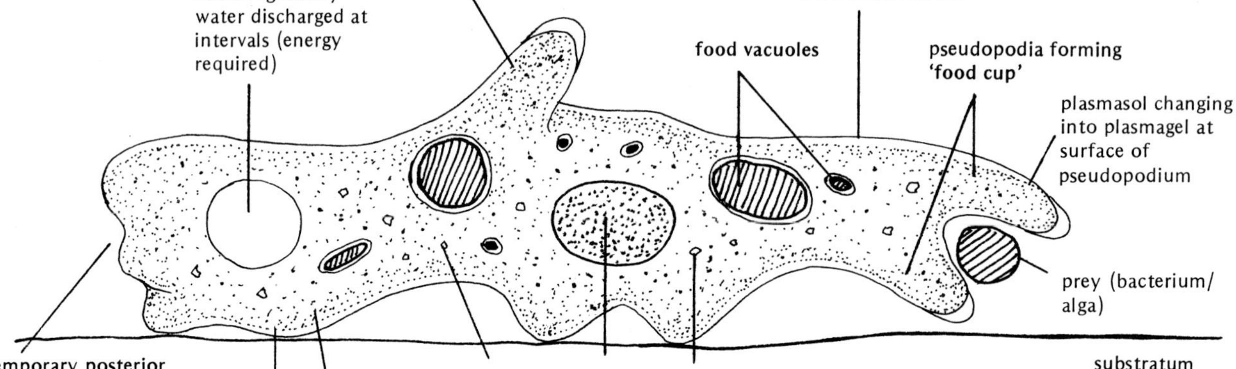

Side view, diagrammatic (Very greatly enlarged. Actual size: up to 500μm diameter)

Notes

1. Amoeba is found at the bottom of ponds, ditches, etc., in damp soil and in the sea.
2. Locomotion, by means of pseudopodia, is slow and irregular, and requires contact with a solid surface. A pseudopodium is formed by the flowing of the protoplasm to produce a blunt projection. This is, of course, accompanied by the withdrawal of protoplasm from other areas. Pseudopodia may form at any point on the surface of the body, and several may develop at once. The size of each pseudopodium is determined by how much endoplasm/plasmasol flows into it (the ectoplasm/plasmagel does not seem to move) and this determines whether the Amoeba moves in that direction. It is not fully understood how the endoplasm is caused to move, or how adhesion to the substratum is achieved, although it is believed that actin and myosin may be involved. Reversible changes in the consistency of the cytoplasm are known to occur, and are probably due to changes in the configuration of the constituent protein molecules (which expand or contract). Possibly at the point where a pseudopodium is forming, the ectoplasm (plasmagel) has been affected by a favourable stimulus (e.g. chemicals or vibrations from food) which causes it to become more liquid. The plasmasol then flows outwards at this point of weakness (due to pulling or pushing forces? – again unknown). As it streams into the developing pseudopodium, any which comes into contact with the environment gels, to form new ectoplasm (with the elimination of water? – there always seems to be a clear region at the tip of an advancing pseudopodium). Conversely, unfavourable stimuli inhibit pseudopod formation. Thus Amoeba moves towards favourable stimuli and away from adverse conditions.
3. Asexual reproduction, by binary fission, occurs when Amoeba has reached a critical size, dependent on its surface area/volume ratio.

CLASS FLAGELLATA

e.g. *Euglena viridis*

Phylum Protozoa
Class Flagellata
Genus *Euglena*

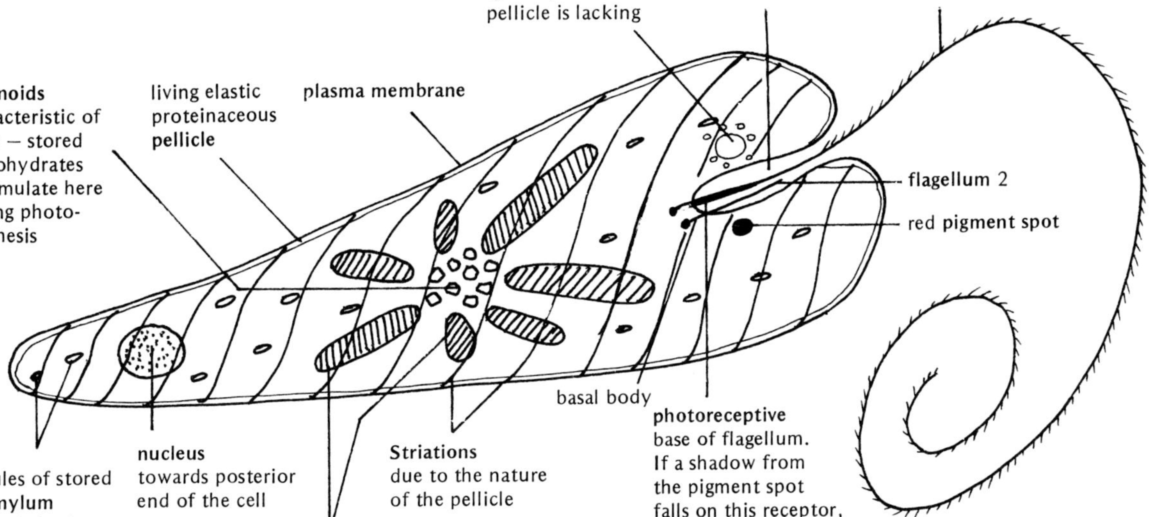

(Very greatly enlarged. Sizes of Euglena range from 50-200μm in length)

Notes

1. Euglena spp. are found in non-flowing fresh water, especially when it contains organic nitrogenous compounds. E.g. In ponds and ditches contaminated with the excretion of farm animals.

2. Euglena can creep over surfaces as well as being able to swim. Creeping involves marked changes in shape, known as euglenoid movements. Swimming is achieved by undulations of the flagellum. Its beating is intrinsic (it will continue lashing when removed from the organism, as long as ATP is available). It is usually held in front or at the side of the swimming cells, like a spinning lasso, causing the cell to rotate and describe a spiral path (i.e. to gyrate) as it is pulled forward through the water.

3. Although capable of photosynthesis, all species of Euglena require some organic materials from the water in which they live, i.e. they are heterotrophic. A few species can live saprophytically in darkness, e.g. *E. gracilis.*

4. The pellicle is not homologous with a plant cell wall; it is internal to the plasma membrane, and is living. In Euglena the pellicle consists of 'strips' wound spirally around the cell, overlapping and articulating at the edges. The striations that result were formerly believed to be contractile fibrils, known as 'myonemes'.

The classification of the Flagellates

The classification of the euglenoids obviously poses difficulties for those who insist that all organisms must be either plants or animals. When living things are classified into only two kingdoms there is no satisfactory way of classifying flagellated unicellular organisms, and we have the ridiculous situation of the same genera being classified into classes of the 'algae' by botanists, and into orders of the class Flagellata of the Protozoa by zoologists.

There is a tremendous diversity of forms and nutrition among flagellated unicells in general. Some (known as Phytomasts) are plant-like in essential respects, being autotrophic and possessing chlorophyll and cellulose walls, e.g. Chlamydomonas, most dinoflagellates, the colonial Pandorina, and Volvox, a colonial form which approaches the multicellular condition. But colourless saprophytic forms (e.g. Polytoma – very similar to Chlamydomonas) occur even in this plant-like group, and many dinoflagellates have a tendency to holozoic nutrition.

Other flagellates (known as Zoomasts) are typically animal-like, and do not possess cell walls or chlorophyll. Some of these feed holozoically, e.g. the Collar-flagellates, and also Trichonympha which lives symbiotically in the gut of termites and ingests the particles of wood eaten by the insect. Trypanosoma (the parasite causing sleeping sickness) absorbs dissolved food through its surface membrane.

The difficulties in classifying these organisms is thus evident, but, despite the existence of atypical forms, it is possible to distinguish obviously animal-like orders from plant-like orders. That is, until we come to the Euglenas and their allies. All species of Euglena have no cell wall but do normally possess chlorophyll and their nutrition can be described as holophytic, although not autotrophic, since all species require some organic materials from their surroundings. In fact, some can live only in water rich in organic nitrogenous matter. Most species cannot survive in prolonged darkness, and in this respect they clearly resemble plants. However, certain species *can* live saprophytically in darkness (e.g. *E. gracilis*) provided there are simple organic nutrients in the water. When kept in prolonged darkness *E. gracilis* becomes colourless, and it is then indistinguishable from Astasia a colourless obligate saprophytic flagellate, in the same order. Peranema, too, belongs to the same order as the Euglenas and is structurally very similar apart from lacking chloroplasts; some members of this genus can ingest solid food particles via the gullet. These species are thus animal-like in every respect.

So here we have a group of closely related organisms that cannot with any justification be excluded from either the plant kingdom or the animal kingdom.

It is probable that the flagellates are an extremely primitive group of organisms ancestral to all other groups, apart from bacteria.

CLASS CILIOPHORA
e.g. *Paramecium sp.*

Phylum Protozoa
Class Ciliophora
Genus *Paramecium*

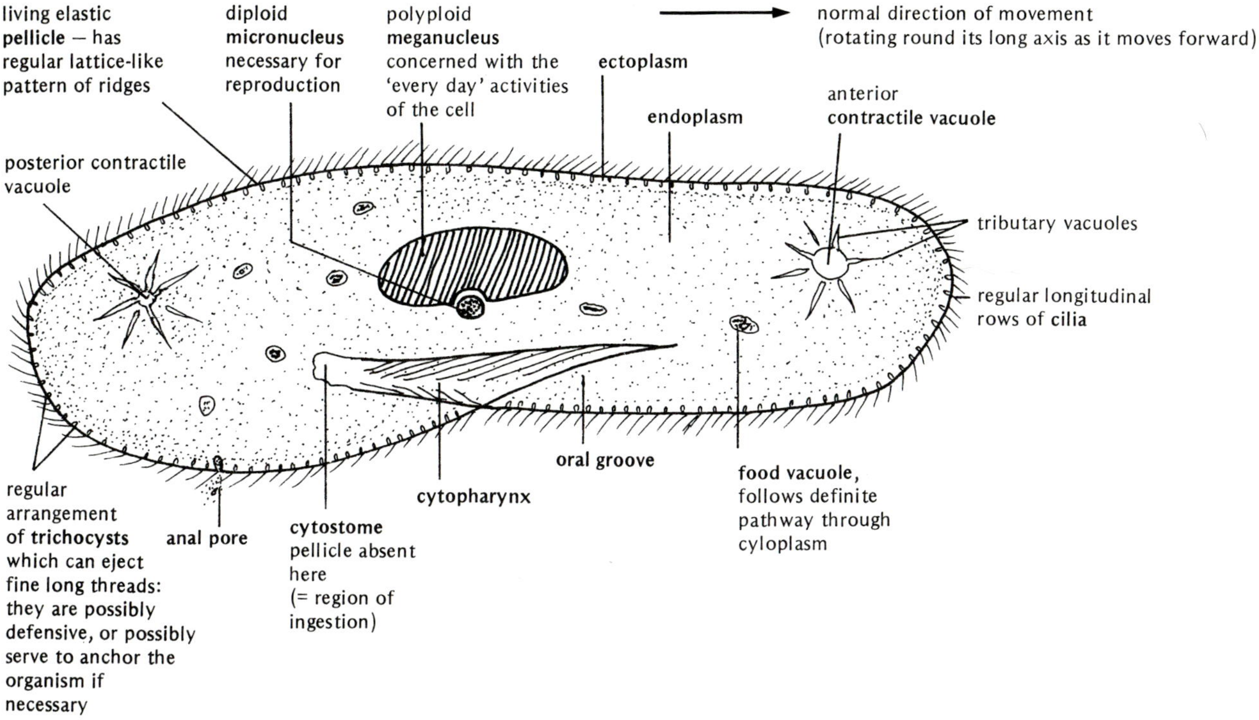

Diagram to show structure (x600. Size: 200μm long)

Notes

1. Found in fresh water containing decaying matter, feeding mainly on bacteria.
2. Can move rapidly by ciliary action – either swimming, or gliding over surfaces.
3. Ciliary action is also effective in directing food particles into the cytopharynx; the food is concentrated into a ball, and then taken up into a food vacuole. Paramecium is able to select which particles it engulfs.
4. Each contractile vacuole discharges through a permanent pore in the pellicle. The posterior vacuole fills and discharges more frequently than the anterior since water passes most rapidly into the body in the region of the cytostome, due to the additional surface exposed here.
5. The cilia are delicate extensions of the ectoplasm. Each contains fibrils arranged in the characteristic 9 + 2 pattern. Each cilium originates from a basal granule; these seem to be linked to longitudinal running fibres which may be concerned with the co-ordination of the beating of the cilia (metachronal rhythm).

Reproduction

Asexual reproduction is achieved by transverse binary fission. The micronucleus divides mitotically, but the meganucleus divides amitotically (i.e. by simple constriction).

Sexual reproduction involves conjugation between differing mating strains of which there are several types. During conjugation the nuclei undergo complicated changes:

Sexual reproduction in Paramecium (Diagrammatic)

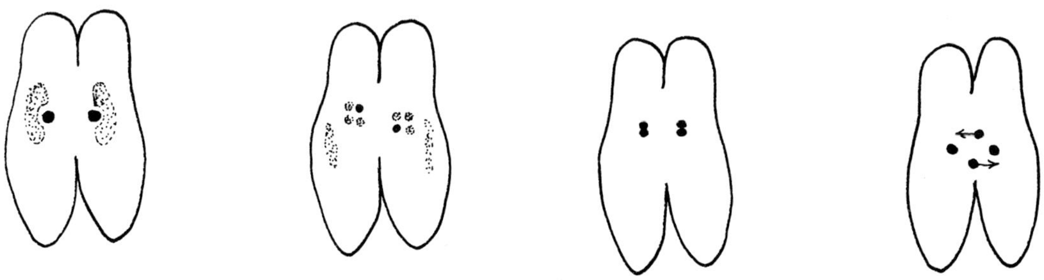

1.
Conjugants of different mating strains unite by their oral surfaces. The meganucleus of each gradually distintegrates and is absorbed into the endoplasm.

2.
The micro nucleus of each divides meiotically, resulting in 4 haploid nuclei in each conjugant. 3 of these abort.

3.
The remaining haploid nucleus in each divides mitotically, producing two gametic nuclei – one migratory and one stationary.

4.
Each migratory nucleus moves into the opposite Paramecium.

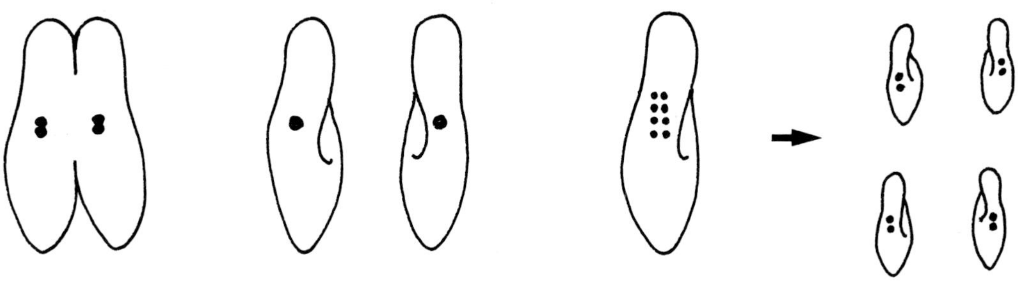

5.
The migratory and stationary nuclei fuse to form a diploid zygotic nucleus.

6.
The conjugants now separate.

7.
The zygotic nucleus of each ex-conjugant then divides several times to produce 8 diploid nuclei.

8.
4 daughter individuals, each with two diploid nuclei, are produced from each ex-conjugant (by binary fission twice). One of the nuclei in each Paramecium enlarges, its genetic material duplicates and it becomes a new polyploid meganucleus.

Notes

1. *Autogamy* may also occur. This is a sexual process, but the gametic nuclei fuse within the *same* individual.
2. Both conjugation and autogamy restore the correct chromosomal balance to the meganucleus. The repeated amitotic divisions of the meganucleus during binary fission result in a culture of Paramecium becoming 'senescent' if neither conjugation nor autogamy take place.

CLASS SPOROZOA

Phylum Protozoa
Class Sporozoa
Genus *Plasmodium*

e.g. *Plasmodium vivax* (malarial parasite)

Life cycle

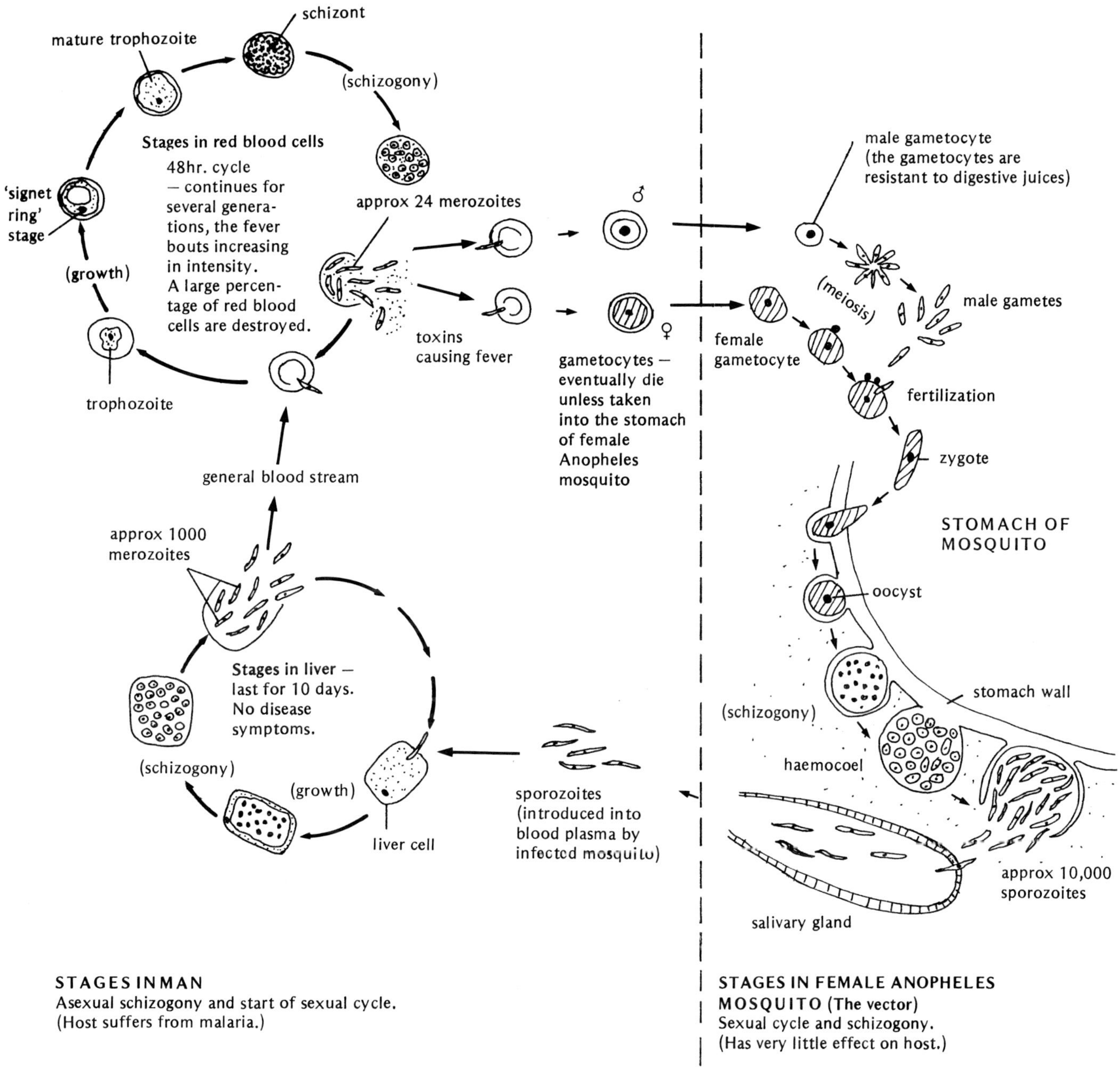

Three forms of malaria can occur, due to different species of plasmodium. In general their life histories are very similar.

Phylum Protozoa
Class Sporozoa
Genus *Monocystis*

e.g. *Monocystis agilis*

A parasite in the seminal vesicles of the earthworm, absorbing the nutrient fluid of the seminal vesicles. (Causes little inconvenience to the host.)

Life cycle

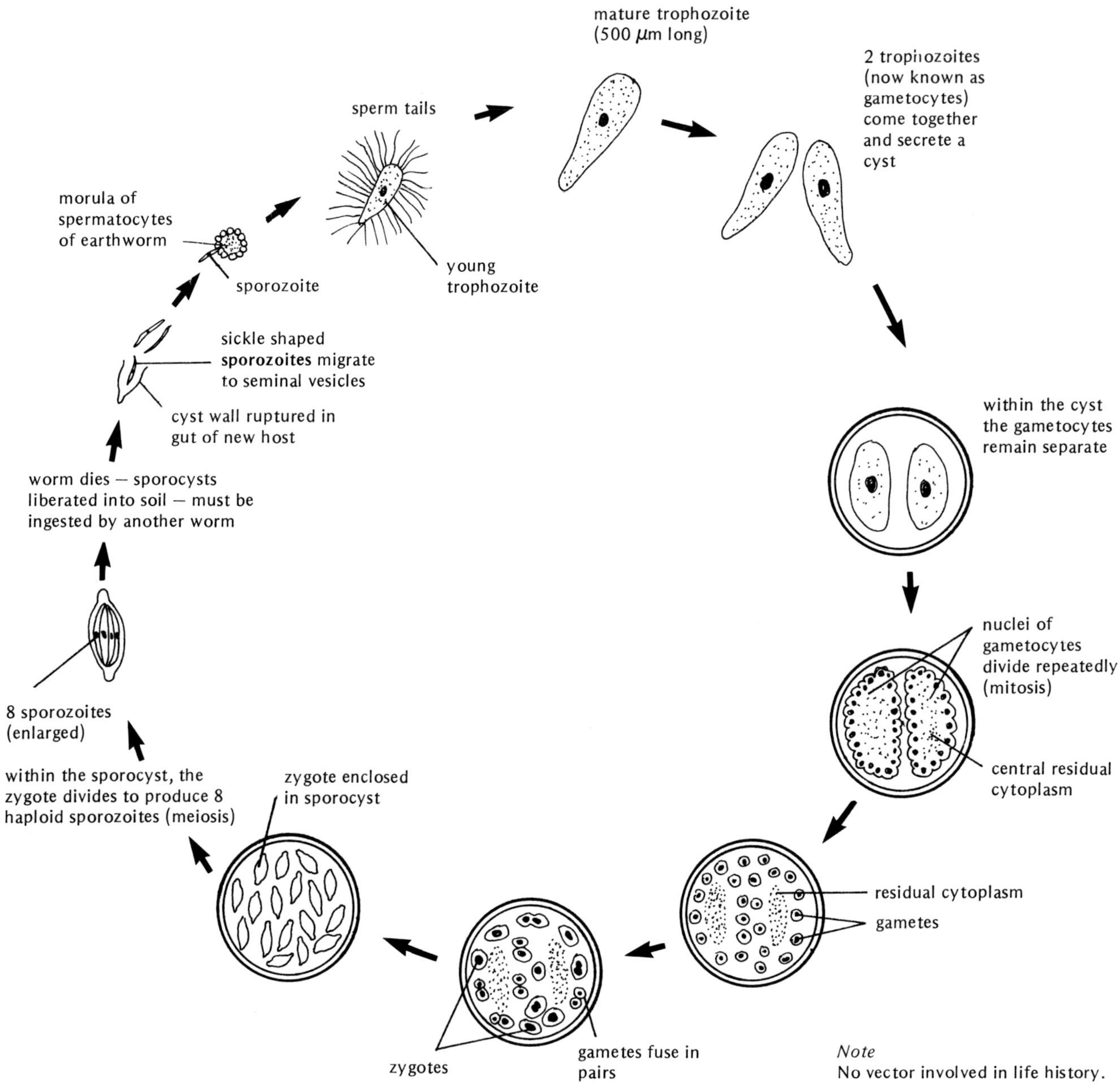

Note
No vector involved in life history.

Various other Protozoa

Amoeboid

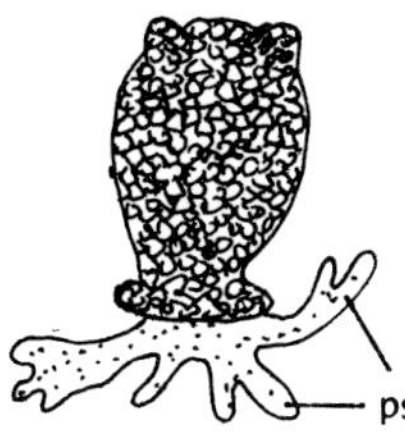

Difflugia (x 100)
Found in ponds

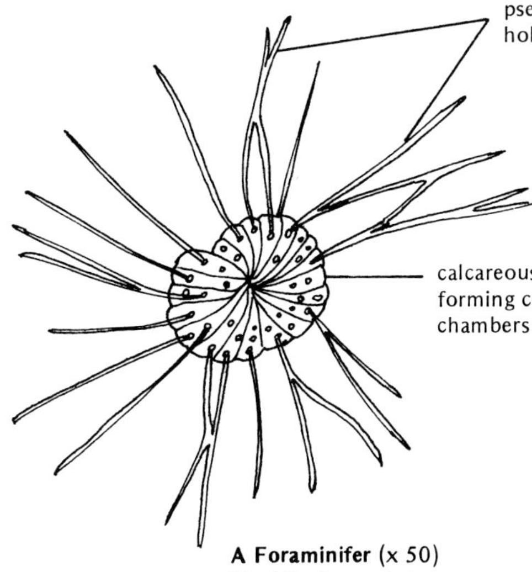

A Foraminifer (x 50)
Planktonic, marine

Flagellated

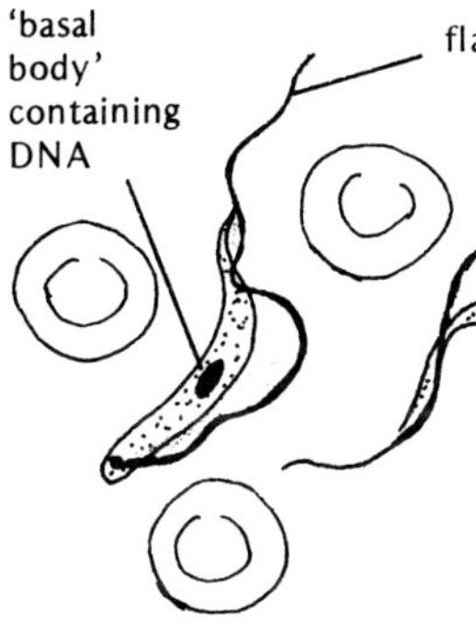

Trypanosoma (x 1000)
In human blood (sleeping sickness)

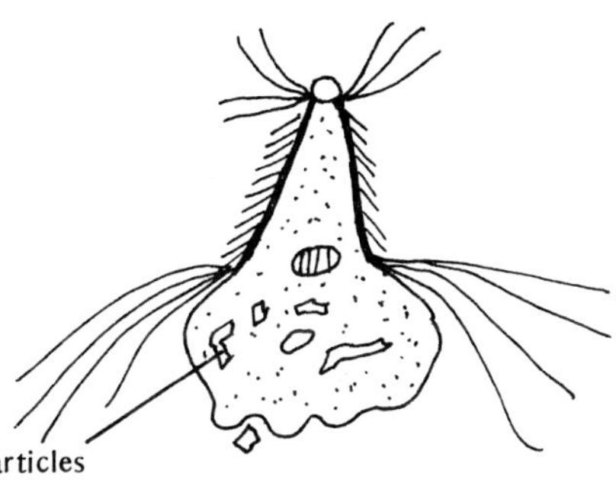

Trichonympha (x 300)
Found in gut of Termites

Codosiga (x 100)
A collar flagellate found in ponds

Cilated

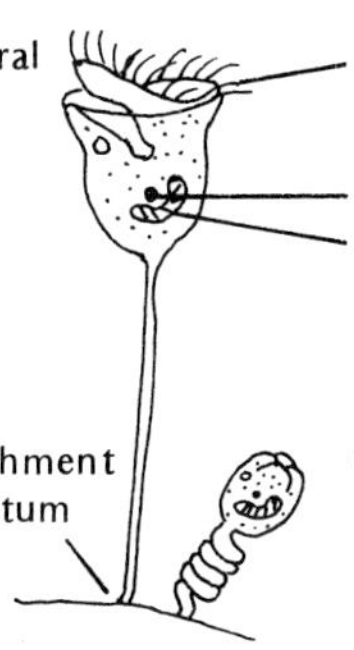

Vorticella (x 100)
Common in ponds.
The 'head' can become
detached and swim away,
having developed a posterior
circlet of cilia.

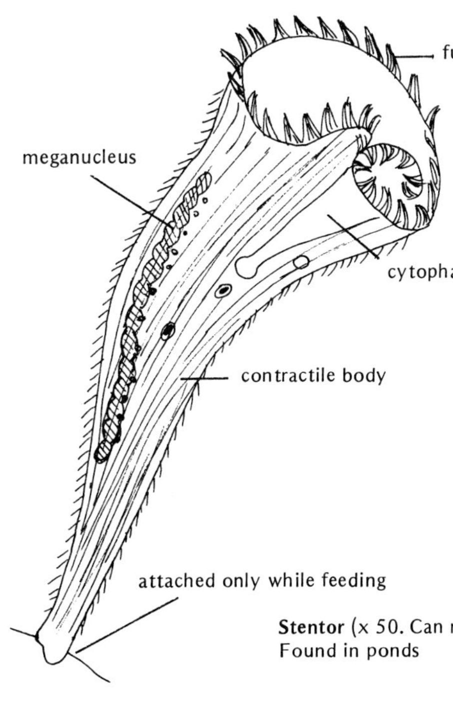

Stentor (x 50. Can reach 3 mm in length)
Found in ponds

6. The Animal Kingdom

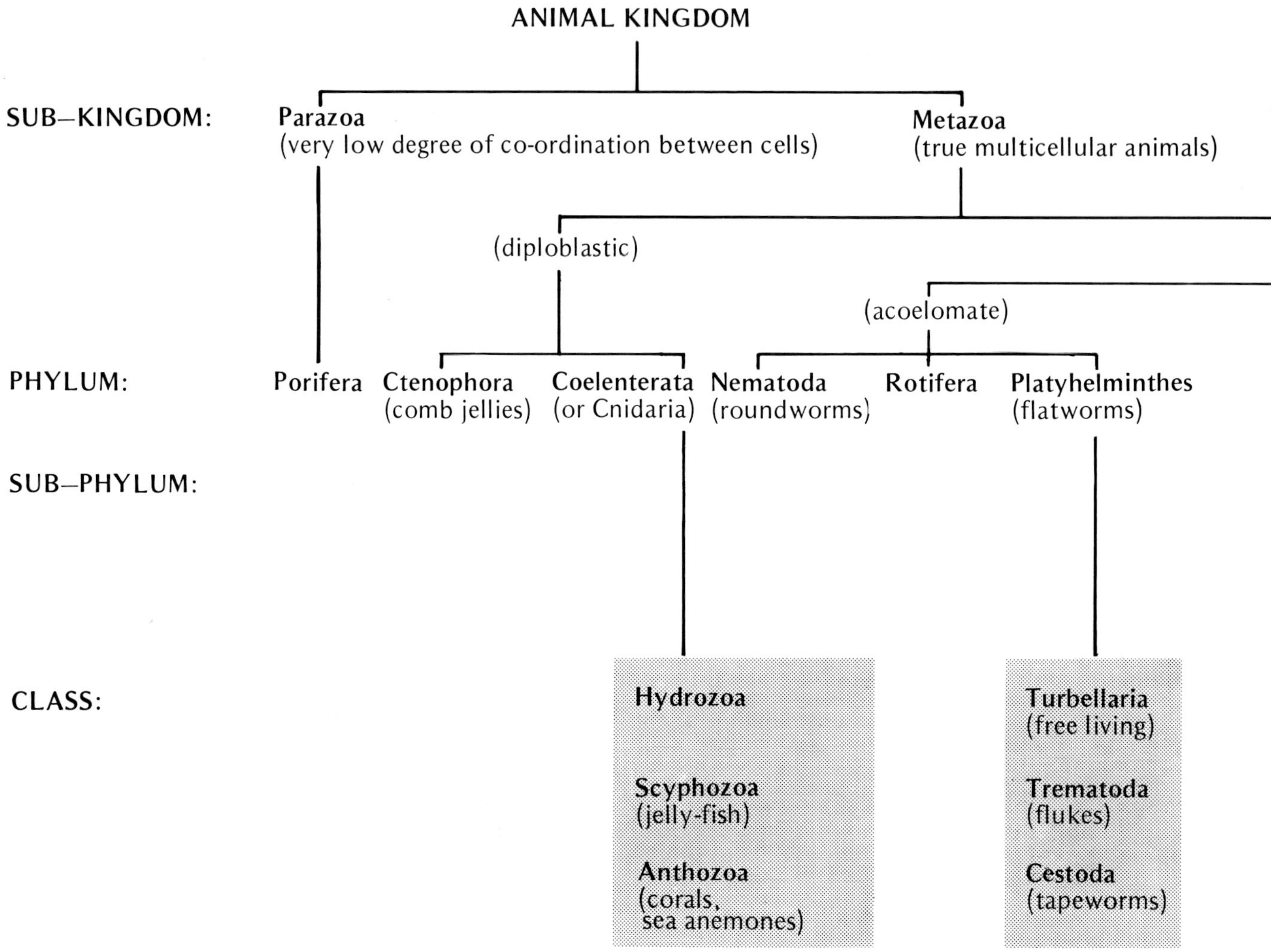

Multicellular animals

As seen in the Protozoa, a single cell can function as a complete individual, since many specialisations are possible within the limits of a single cell. However, unicellular organisms can never attain a large size since one nucleus can exert control over only a certain volume of protoplasm, and also a large cell has a relatively small surface area compared with its volume (unless there are extensive infoldings of its surface). A *high* surface-area/volume ratio is required by a mass of protoplasm if the diffusion of oxygen, metabolites and waste products is to occur fast enough to support life.

There are also restrictions to the extent of differentiation that is possible within a single cell. So a unicell is limited not only in size, but also in degree of specialisation, and this reduces the range of habitats open to such organisms. Unicellular organisms are very much at the mercy of their environment.

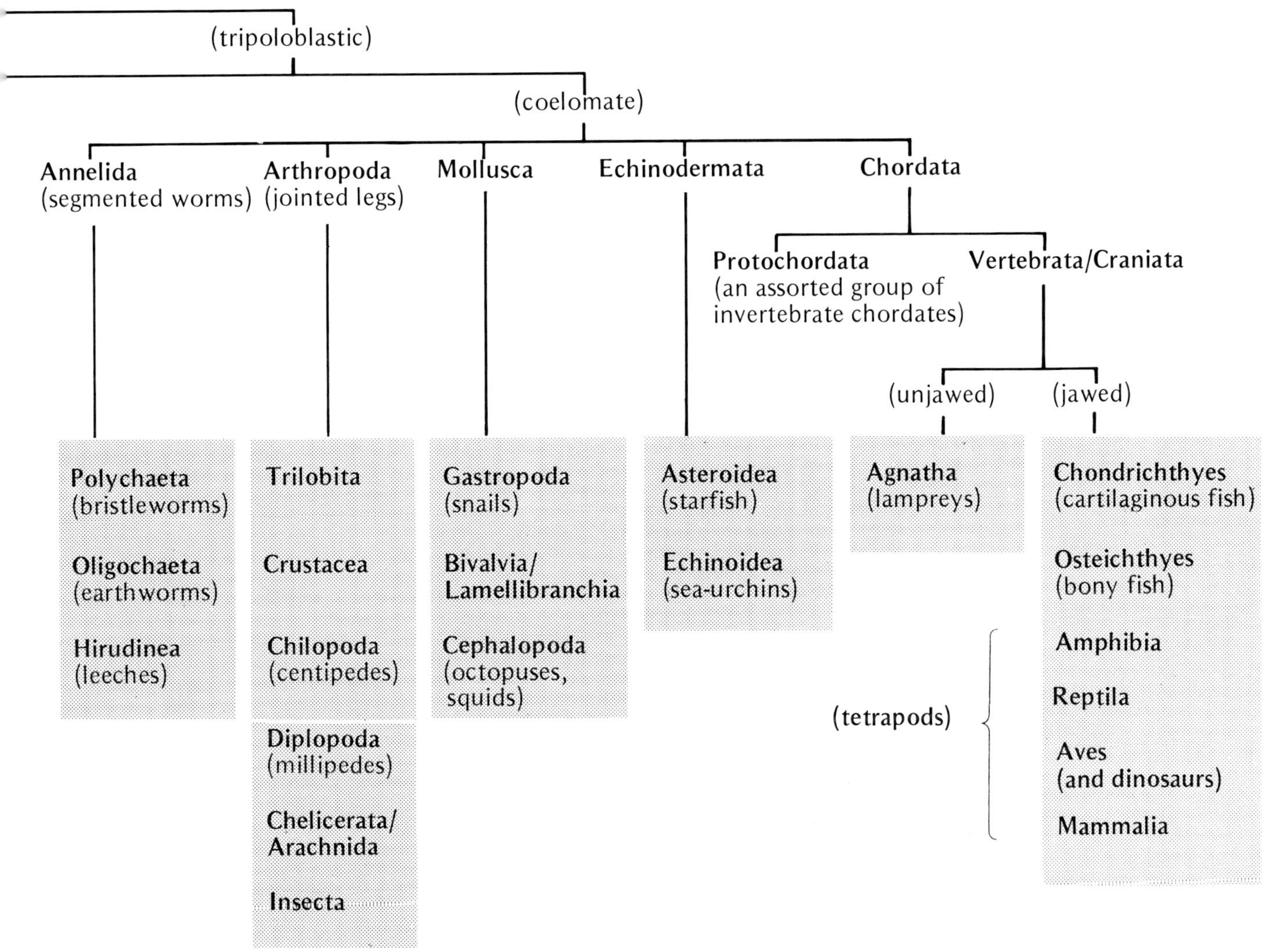

The division of a large mass of protoplasm into discrete nucleated cells makes it possible for them to have spaces between them, and thus overcomes the surface/volume problems. It also enables a large body of protoplasm to support itself to a greater extent and maintain a constant shape, i.e. the stiffening result of partitioning the protoplasm into microscopic cells, each enclosed in a membrane, can be compared with stiffly beaten egg white as opposed to raw egg white.

Multicellularity also opens up the possibility of different cells becoming specialised for different purposes, only some cells feeding while others are specialised for protection, locomotion, support, reproduction, etc. Such division of labour depends on co-ordination between the differing cells, which are now dependent on each other for survival. The drawback in this respect is far outweighed by the benefits derived from the multicellular state; it permits much greater functional efficiency and considerable diversity of form and complexity of behaviour. It has enabled different multicellular organisms to exploit a vast range of habitats, and to achieve an increasing degree of mastery over their environment as evolution has progressed, i.e. in the more advanced forms.

Thus a multicellular organism can be described as an individual composed of many different specialised interdependent cells, a definition which seems perfectly straightforward. However, it is occasionally impossible to decide whether members of certain species exist as colonies of individual unicells rather than single individual multicellular organisms. Volvox provides the classic example. Volvox takes the form of a hollow sphere composed of several thousand Chlamydomonas-like cells embedded in jelly-like material. There is a surprising degree of co-ordination between the cells, which are connected by very fine protoplasmic strands. A particular end of the sphere is always directed forwards during swimming, which indicates that the beating of the flagella of the cells is co-ordinated. The cells at the anterior end have enlarged 'eye spots'. In the rear of the sphere larger 'reproductive cells' can be found; only the cells in this part of the sphere are capable of reproduction. If one of the vegetative cells becomes isolated from the sphere it cannot live independently and soon dies. During sexual reproduction in Volvox fully differentiated gametes are formed: egg cells by the enlargement of certain cells as they become loaded with food, and numerous small flagellated sperm by the repeated divisions of other cells in the same or another colony. During the development of a new colony, either sexually or asexually, cell divisions occur synchronously. Thus specialisation of cells, co-ordination, and division of labour occur in Volvox, but only to such a limited extent that Volvox is not usually considered truly multicellular. The term coenobium is used to describe such colonies of cells, i.e. a colony where a constant number of cells is arranged in a specific manner and co-ordinated in such a way that the colony behaves as a unit.

Other members of the Volvocales, e.g. Gonium, Pandorina, form similar colonies, but the numbers of individual cells involved is smaller, there is no protoplasmic continuity between them, and there is no division of labour. If any cell becomes isolated from the colony it is capable of swimming away and dividing to form a new colony, i.e. in these cases the colony is merely a cluster of independent cells.

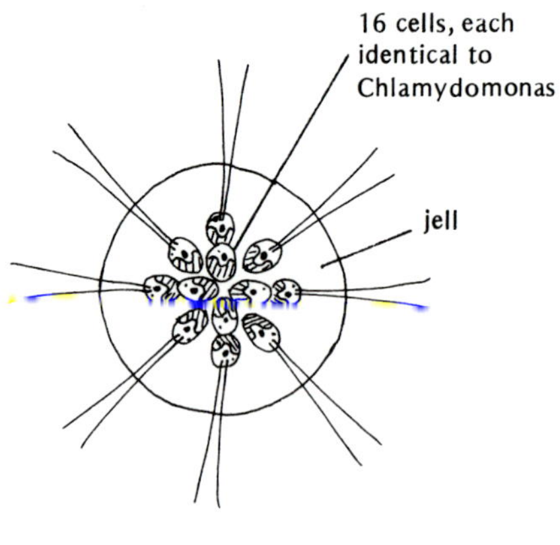

Pandorina (Spherical or oval)

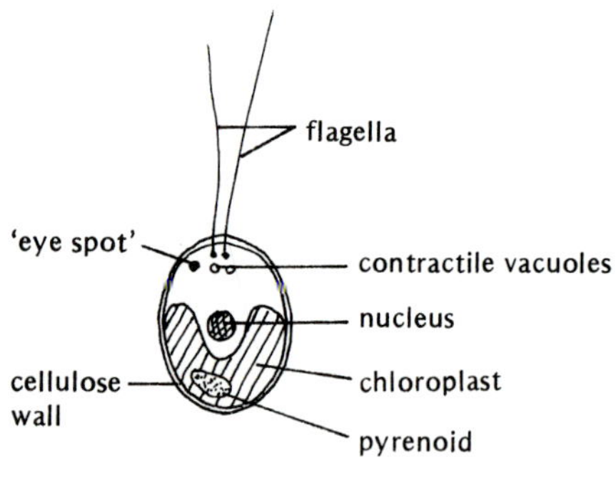

Chlamydomonas (20 μm)

Thus the line between colonies of individual cells and a single multicellular organism is not always easy to draw. The sponges are another group which illustrates this point; the degree of cell differentiation and of intercellular co-ordination is insufficient to justify their inclusion in the Metazoa.

Sub Kingdom Parazoa

7. Phylum Porifera (sponges)

There is only a single phylum in this sub-kingdom.

Introduction

Although sponges can be considered multicellular animals, their body plan is unique. It is believed that their origin (they probably evolved from collar flagellates) is distinct from all other multicellular animals and therefore, although their body structure is similar in certain respects to Coelenterates, different names are applied to the analagous parts, and sponges are not included in the Metazoa (true multicellular animals). They are usually described as 'cell-aggregates'.

General characteristics

Aquatic, mostly marine – sessile, permanently attached and non-mobile – usually colonial (distinct individuals are not clearly delimited).

Abundant and widespread, especially in the sea.

Sac-like body consisting of essentially two layers of cells surrounding a gastric cavity, but considerable folding of the body wall usually obscures the basic sac-like form.

Body wall is perforated by numerous intracellular pores through which the feeding current of water is drawn *into* the gastric cavity.

Water passes out of the gastric cavity via large osculi.

Cell differentiation occurs, i.e. different cells are specialised for particular functions. But individual cells behave remarkably independently and a sponge can be considered a loosely organised aggregation of cells rather than a true multicellular animal. De-differentiation can occur, i.e. a specialised cell can revert to a general form and then re-differentiate into a different type of cell.

No specific tissues or organs. No nervous system – a sponge is very unresponsive.

Gastric cavity lined by flagellated collar cells (choanocytes), not found in any other multicellular animals – responsible for producing the water current and for capturing food.

Body supported by a skeleton of spicules of silica or calcium carbonate, or fibres of horny spongin.

Symbiont algae often present, therefore sponges are often coloured.

Asexual reproduction by a variety of means, e.g. budding, branching and fragmentations. Great powers of regeneration. Asexual reproductive bodies (gemmules) are produced by fresh water sponges to survive unfavourable conditions. Larvae can sometimes be produced asexually.

Sexual reproduction occurs. Individuals are hermaphrodite. The embryo develops within the body wall of the sponge. It becomes a free-swimming ciliated larva which disperses the species. The type of embryonic development is unique.

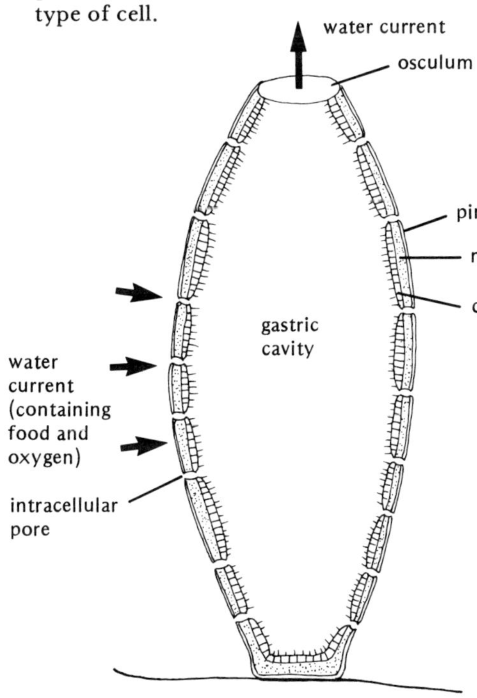

Diagram to show the general structure of a sponge (Greatly simplified)

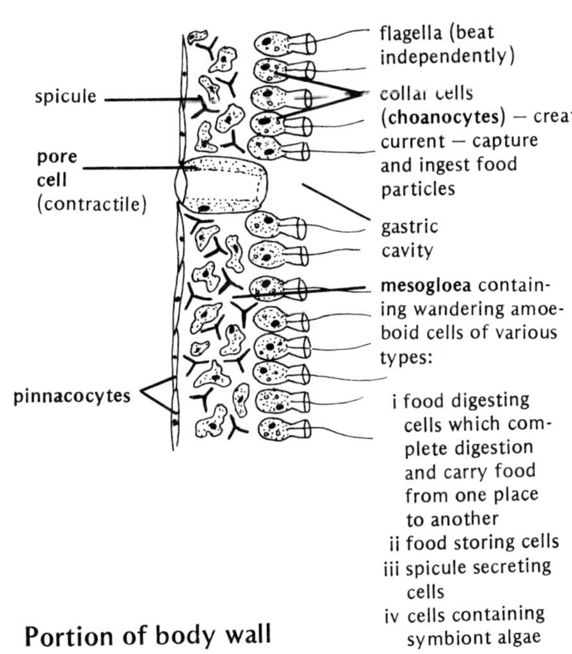

Portion of body wall

Sub Kingdom Metazoa

8. Phylum Coelenterata

Introduction

Coelenterates are the simplest of the Metazoa. Although many different types of cell are present (e.g. musculo-epithelial, sensory, glandular, nerve, digestive, sex cells, etc.) these are integrated to form tissues at a more primitive level than in other Metazoa, and no structures that can merit the name of 'organ' are present. However, since a nervous system is present, there is definite co-ordination between cells and an individual animal can react as an entity. The behaviour of Coelenterates, although simple, is therefore far more varied and complex than that of sponges.

Coelenterates are *diploblastic*. This means that the body is composed of only two layers of cells, ectoderm and endoderm, i.e. all the cells in the body are ectodermal or endodermal in origin. The body is basically radially symmetrical and sac-like, and since co-ordination is effected by a diffuse 'nerve net' locomotion and sensitivity are never well developed. Even large jelly fish have very limited powers of swimming, merely sufficient to maintain themselves in the surface waters. They tend to drift with currents.

Individual coelenterates exist in two basic types:

i. Motile solitary free-living *medusae* (jelly fish) which are concerned in sexual reproduction and in dispersing the species.

ii. Attached *polyps/hydroids*, which may be solitary or colonial.

But, in fact, even hydroids and medusae are constructed on the same general plan:

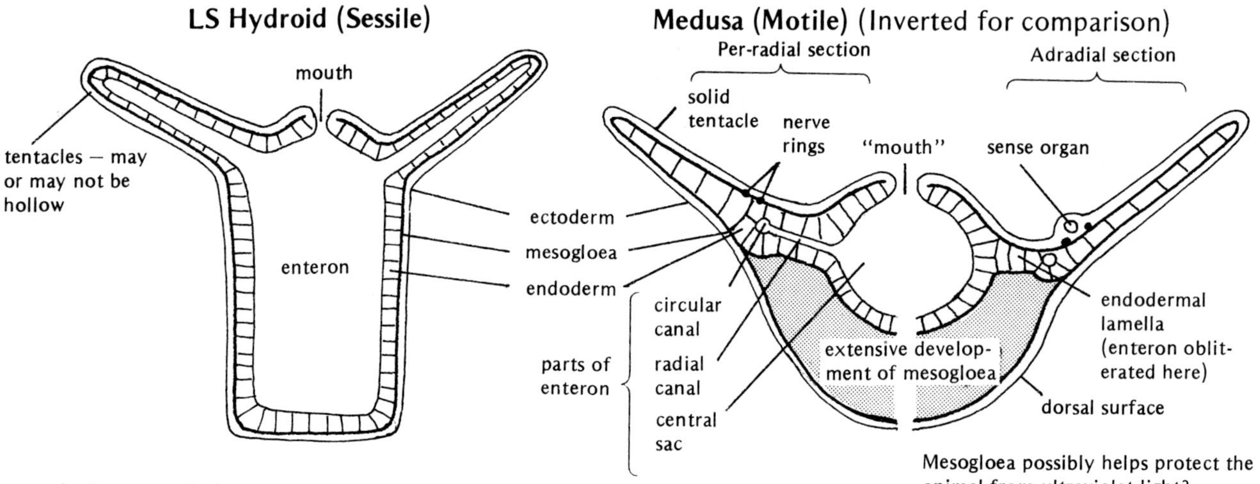

General characteristics

All aquatic, mostly marine – often colonial.

Abundant and widespread in the sea.

Passive way of life.

Diploblastic – radially symmetrical.

Single internal cavity (gastro-vascular cavity/coelenteron/enteron) with only one opening to the exterior, the 'mouth'.

Ring of mobile tentacles surrounding mouth.

Nerve net – no CNS.

Polymorphism common (i.e. the fully developed adult individuals of a single species can exist in various forms). The different types of individual can all be interpreted as variants of a hydroid or medusoid form.

Prey captured by means of nematocysts (stinging capsules) and conveyed to the mouth by tentacles. Coelenterates feed only on living prey.

Symbiont algae usually present in endoderm.

Asexual reproduction confined to hydroids – occurs by budding and branching. Great powers of regeneration.

Sexual reproduction: Coelenterates are usually dioecious (i.e. sexes separate). Typically the gonads develop in the medusae – gametes are liberated into the sea – fertilization is external – fertilized eggs develop into ciliated larvae (planula larvae) which eventually develop into new hydroids. (But no medusae are present in Anthozoa or in freshwater Hydrozoa. Here the gonads are produced by the polyps.)

Class Hydrozoa	Class Anthozoa	Class Scyphozoa
Mostly marine – few fresh water	All marine	All marine
Typically both hydroid and medusoid forms occur in life cycle	No medusoid stage	Reduced hydroid stage in life cycle
Both small and fragile	Complexly organised polyp, with vertical partitions in enteron, and gullet lined by ectoderm	Medusa may reach very large size – extensive system of radial canals – 4 gastric pouches off central sac – relatively well developed sense organs
Medusa possesses velum*		No velum*
Usually colonial	Solitary and colonial forms	Solitary
Many secrete a chitinous skeleton (perisarc)	Colonial forms secrete calcareous skeleton	No skeleton
Gonads ectodermal	Gonads endodermal	Gonads endodermal
E.g. the following genera:		
Obelia **Hydra** **Physalia** (Portuguese man-of-war)	**Actinia** (sea anemone) Also **corals**	**Aurelia** (jelly fish)

Nematocysts (stinging capsules)

These are highly complex organelles produced by cells known as nematoblasts (or cnidoblasts). They are found only in Coelenterates, and are responsible for the success of the phylum. Nematoblasts are extremely specialised ectodermal cells concerned in the capture of prey, and they enable a simple animal, fixed to one spot and lacking complex sense organs or any means for rapid movement, to catch advanced, relatively large and fast moving prey. A nematoblast originates from a 'replacement' cell, and this actively migrates (often via the enteron) while the nematocyst (stinging capsule) is developing within it. Eventually most occur in the tentacles where they are found in enormous numbers, in 'batteries' between and embedded in the musculo-epithelial cells of the ectoderm. To reach their final destination they may sometimes have been ingested by the endodermal cells of the tentacle and then passed from cell to cell.

The nematocyst is a chitinous capsule containing a hollow coiled thread which will automatically shoot out when the nematoblast is suitably stimulated, either by touch or by certain chemicals.
Depending on the type of nematocyst, the thread will either:

i. coil around part of the prey, *or*
ii. adhere to the prey (or to the substratum during locomotion), *or*
iii. inject a powerful paralysing poison into the prey; these threads are ejected with such force that they can penetrate the external skeleton of the prey.

The mechanism of discharge is not definitely known; it is believed that an uptake of water occurs and that pressure changes are involved. The discharge is purely mechanical and does *not* involve the nervous system.

Nematocysts can be used only once. They are gradually extruded once they have 'fired', and must be continually replaced by new ones.

Once the prey is caught, sensory cells are stimulated by its struggling, and a chain of nervous reactions comes into play which result in the prey being ingested via the mouth.

* The velum is an ectodermal membranous circular shelf projecting inwards from the lower margin of the medusa. During development the membrane extends right across the sub-umbrella surface of the medusa, i.e. closing off its under surface.

Nematoblasts of Hydra

Volvent (adhesion)

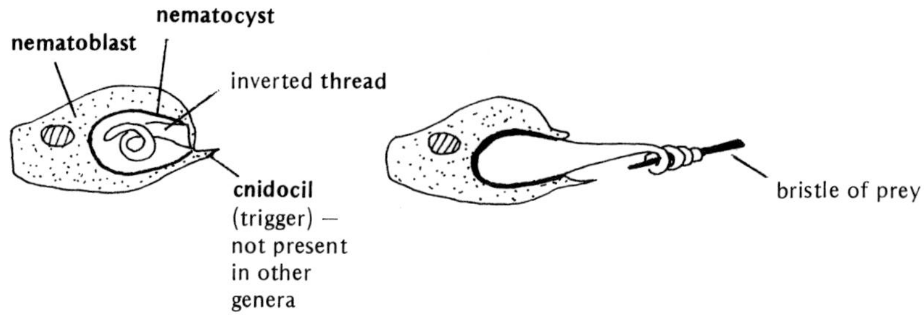

Penetrant (injects poison) — **largest type**

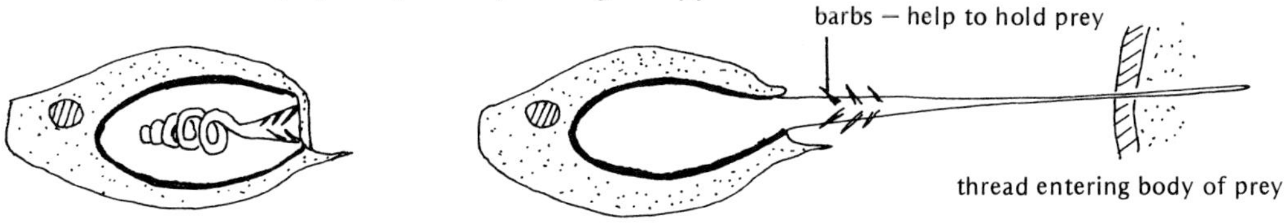

Glutinant (adhesion — used also in locomotion)

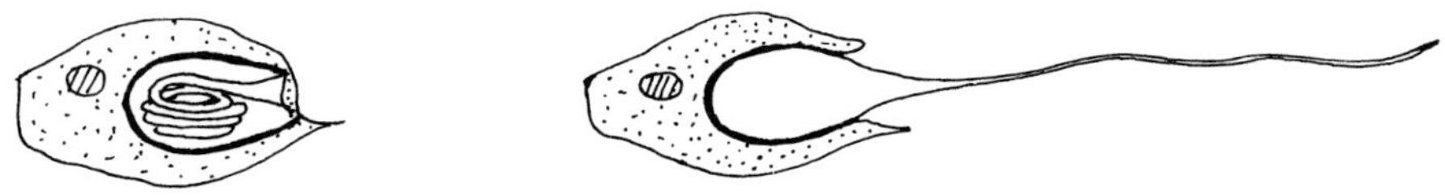

Undischarged **Discharged**

The nervous system of Coelenterates (nerve net)

The nervous system consists entirely of a network of neurons. There is no CNS. Receptors and effectors are connected up in an apparently haphazard manner by a system of multipolar nerve cells, forming a nerve net in which the synapses allow the passage of impulses in either direction. When a sensory cell is stimulated, it evokes impulses which are transmitted by the nerve cells slowly in all directions from the point of stimulation. Since there is no CNS, there are no distinct sensory and motor nerves and no obvious pathways are followed. However the nerve cells are *not* linked haphazardly and the impulses are eventually transmitted to appropriate effectors, which then respond. The responses become weaker further from the point of stimulation. This is because the first impulses to arrive at any synapse do not usually succeed in stimulating the adjacent nerve cell. They do, however, cause an increase in the responsiveness of the next nerve cell so that subsequent impulses arriving at the synapse immediately afterwards *do* pass across. A strong or prolonged stimulus will therefore evoke a more widespread reaction.

The rapid withdrawal response of a polyp is believed to be due to a system of especially large bi-polar nerve cells which are linked to all the longitudinal (ectodermal) muscle tails. These therefore contract almost simultaneously in response to a strong unfavourable stimulus, causing the polyp to contract rapidly.

In the medusae, which are adapted to a motile way of life, the neurosensory system is better developed and the nerve net is more concentrated in certain parts of the zooid. Two nerve rings occur in association with the enlarged muscle tails concerned with the pulsating of the 'bell', and with the sense organs responsible for its orientation.

CLASS HYDROZOA

Phylum Coelenterata
Class Hydrozoa
Genus *Hydra*

A freshwater coelenterate found in ponds.

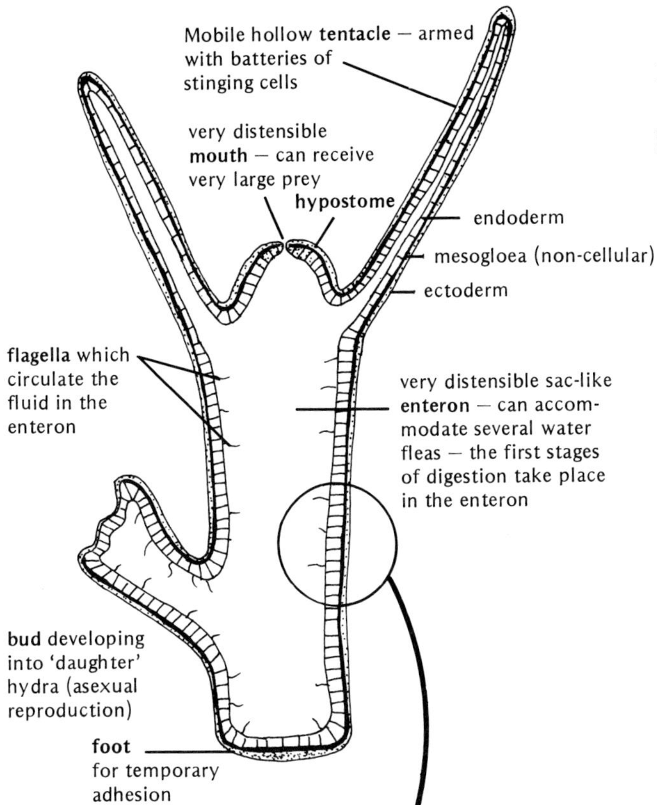

LS Hydra

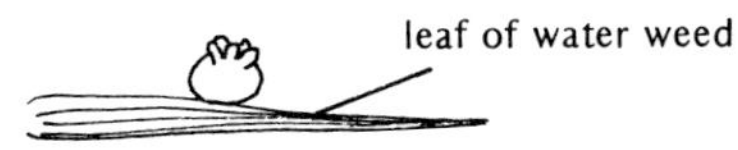

Hydra contracted

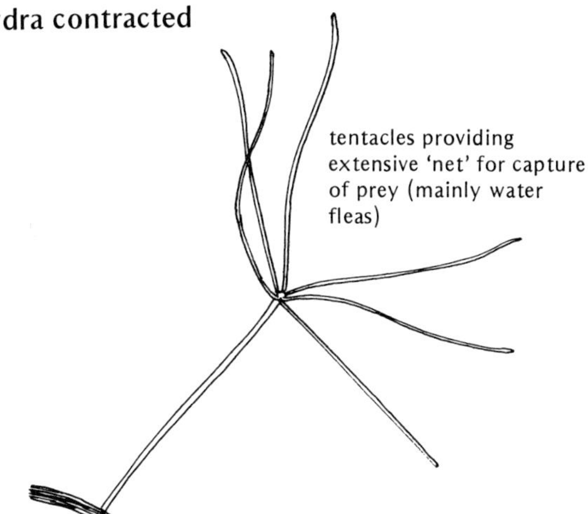

Hydra extended (Approx. x10)

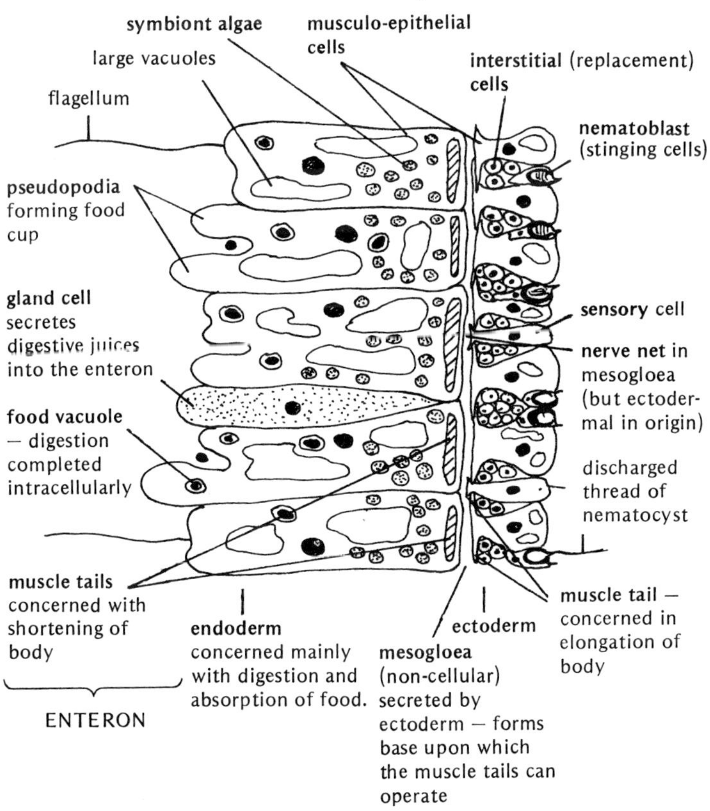

LS portion of body wall of Hydra to show cellular structure (Diagrammatic)

Notes

1. Slow locomotion is possible by looping or somersaulting.
2. Osmo-regulatory mechanism possibly involves the active transport of inorganic ions across the body wall.
3. Most species of Hydra are dioecious. In those that are hermaphrodite the testes develop before the ovaries, and occur nearer the mouth end of the body.

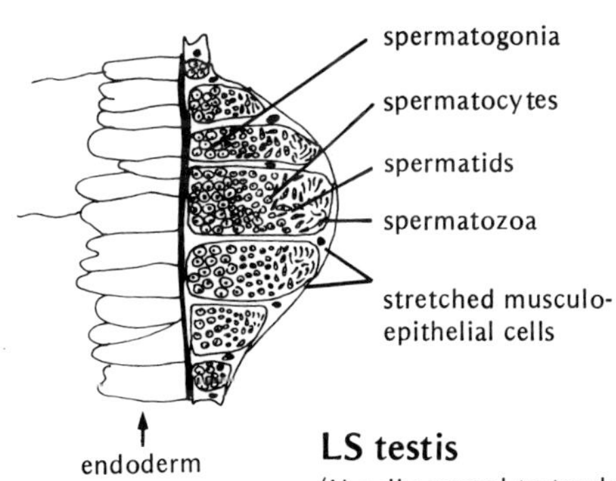

LS testis
(Usually several testes develop simultaneously)

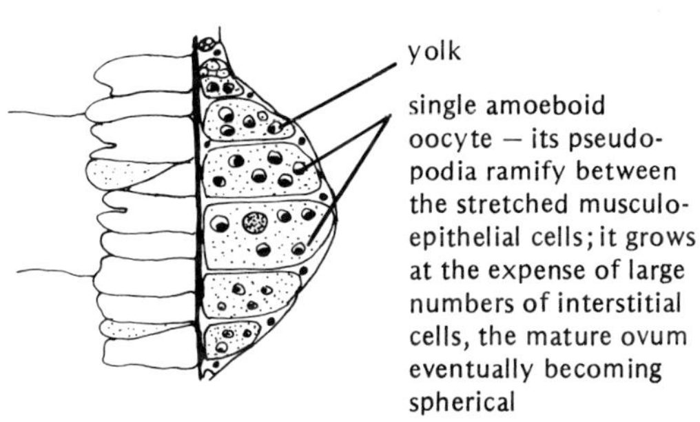

LS developing ovary
The gonads are temporary structures – develop when conditions are becoming unfavourable.

Phylum Coelenterata
Class Hydrozoa
Genus *Obelia*

e.g. *Obelia geniculata*

A marine hydroid colony, attached to seaweeds at low tide level and in deep water.

Appearance of colony (x4)

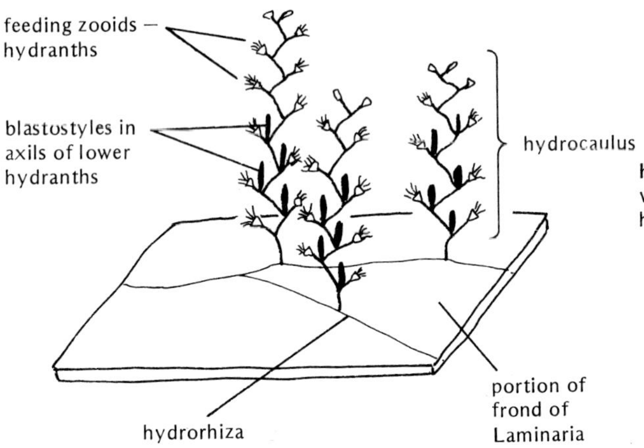

Prey consists of larvae and various minute organisms of the plankton

Detail of Hydrocaulus (Greatly enlarged)

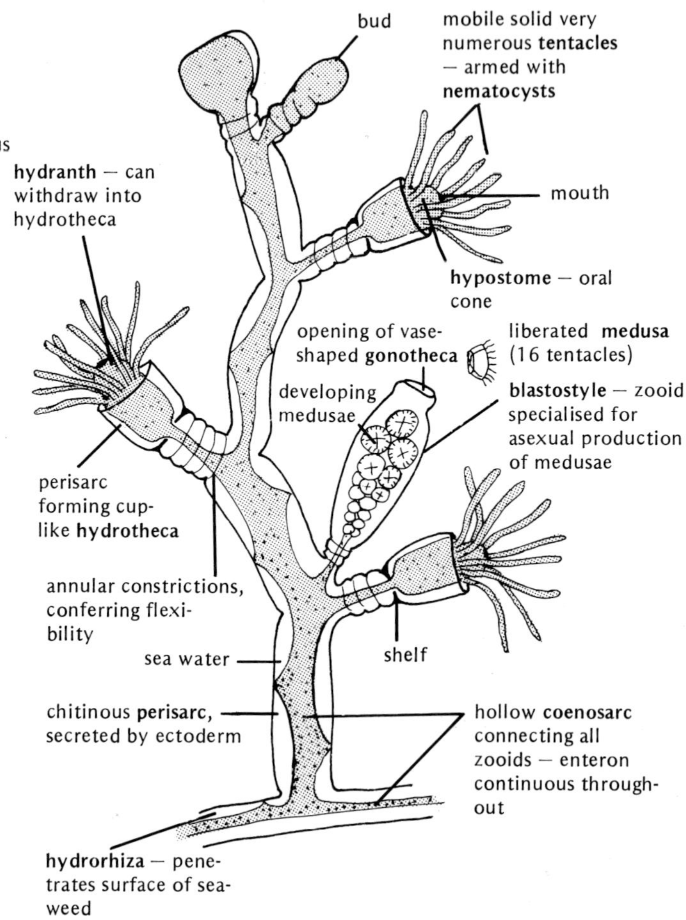

Diagram of Medusa to show structure

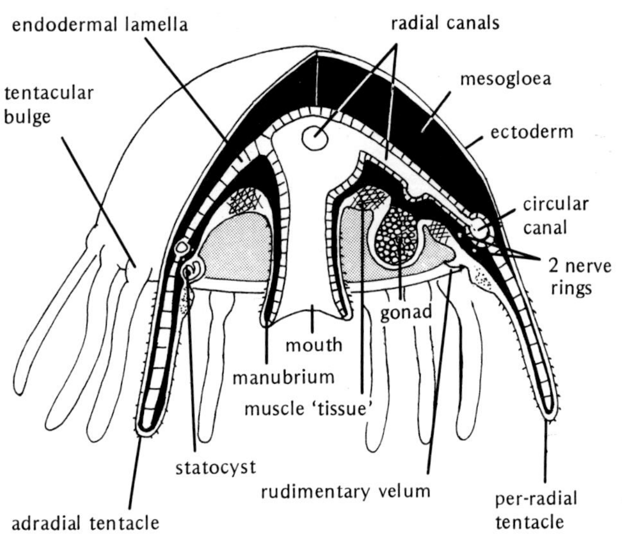

Medusa from ventral surface (Greatly enlarged)

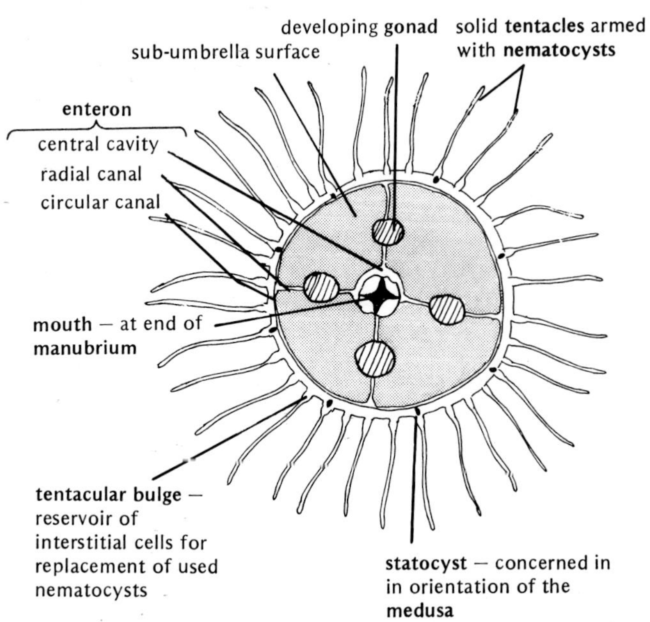

Notes

1. A medusa is either male or female.
2. It is a free-living zooid.
3. A medusa is very small when first liberated, but grows to ½cm in diameter when mature.
4. The gonads burst when ripe and the gametes are liberated into the sea.

Life cycle of Hydra

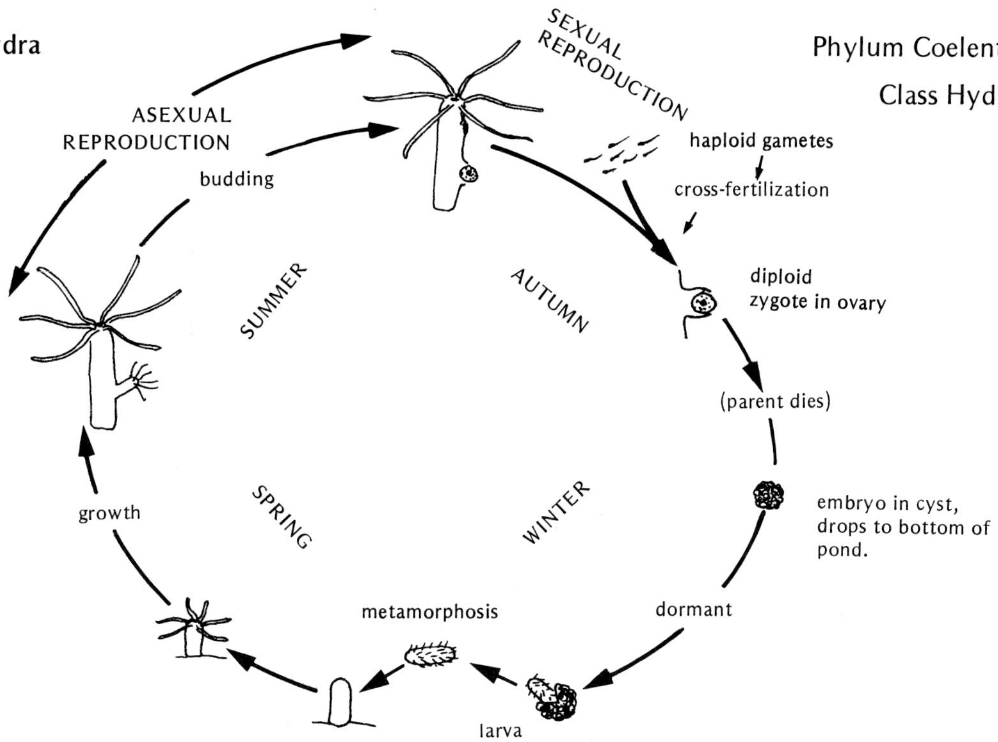

Notes

1. The absence of a medusa in the life cycle is a modification associated with life in a pond.
2. In hermaphrodite species the testes develop before the ovary/ies (i.e. protandrous). They are situated towards the mouth end of the body.

Life cycle of Obelia

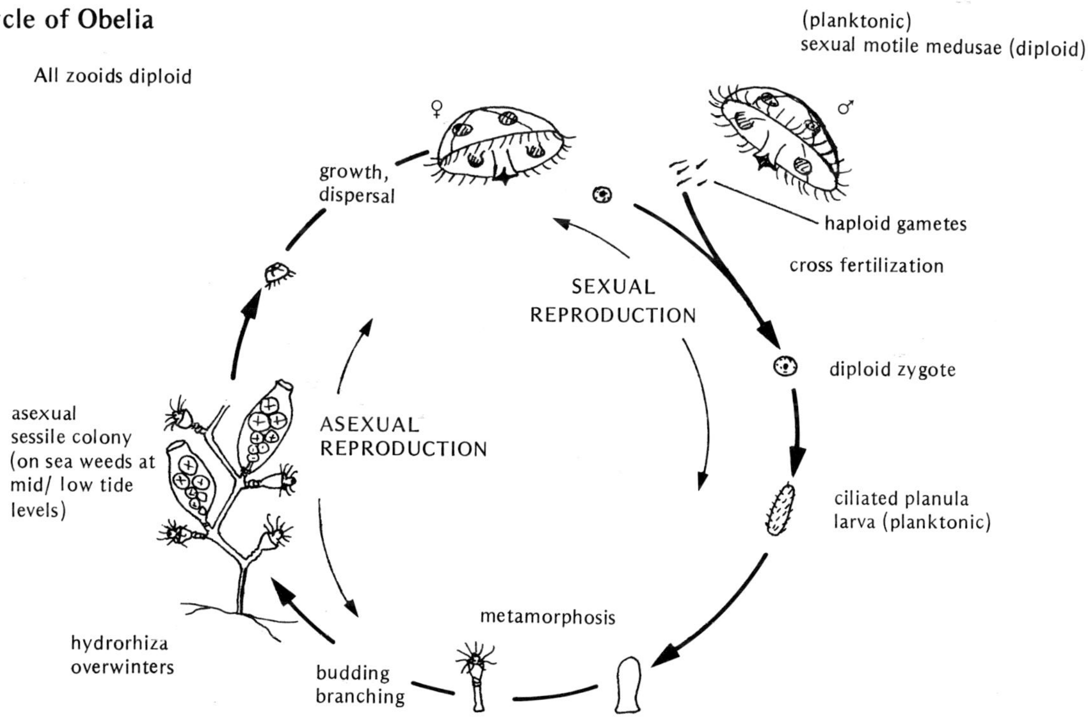

Notes

1. The life cycle of Obelia is extremely hazardous. The dispersal of offspring to new locations is a difficult problem, since the colony is fixed. The production of motile medusae helps in this respect and assists in cross-fertilization.
2. The asexual reproduction of the colony ensures that a single successful offspring eventually gives rise to very many medusae, to compensate for the considerable losses.

Phylum Coelenterata

Class Scyphozoa

Genus *Aurelia*

CLASS SCYPHOZOA

e.g. *Aurelia aurita* (common jelly-fish)

Life cycle

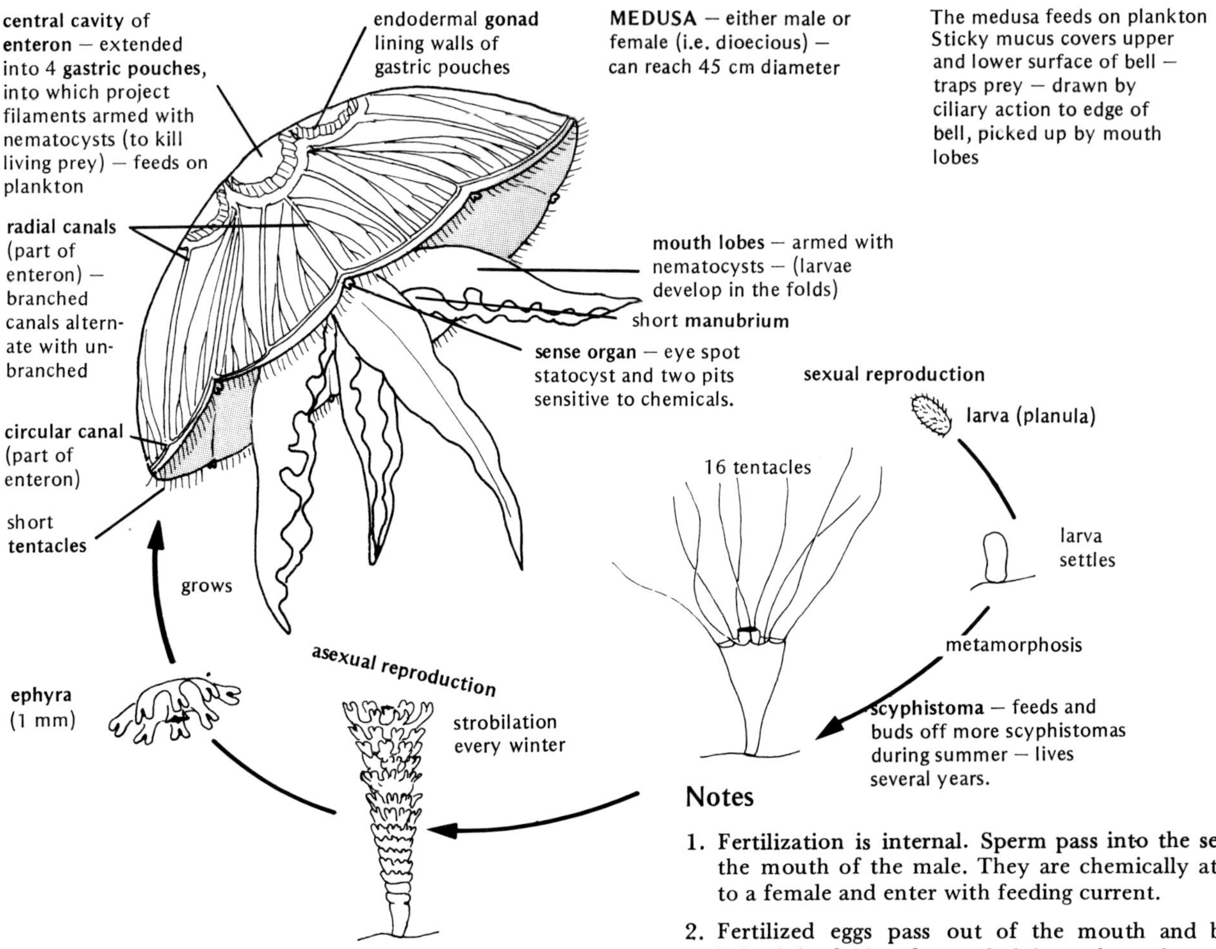

Notes

1. Fertilization is internal. Sperm pass into the sea from the mouth of the male. They are chemically attracted to a female and enter with feeding current.
2. Fertilized eggs pass out of the mouth and become lodged in folds of mouth lobes where they develop into larvae.

Class Anthozoa

Genus *Actinia*

CLASS ANTHOZOA

e.g. *Actinia equina* (sea anemone)

Notes

1. **Symbiont algae** are present.
2. There is no asexual reproduction.
3. Internal fertilization takes place. Baby polyps escape via the mouth (**viviparous**).

Cut away to show structure (x2)

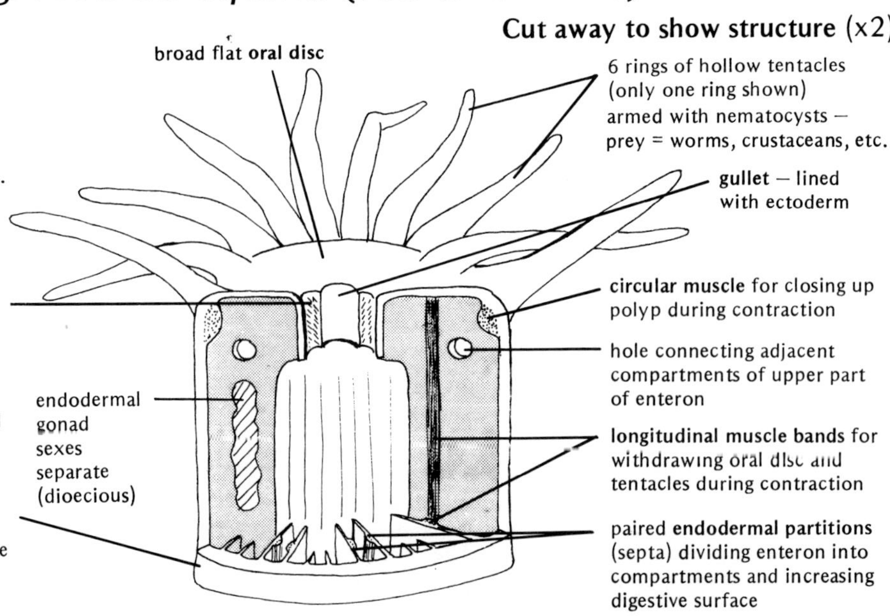

9. Phylum Platyhelminthes (flatworms)

Introduction

All Metazoa other than the Coelenterates are triploblastic, since in the embryo a third layer of cells can be distinguished between the endoderm and ectoderm, i.e. the mesoderm. These three layers usually become indistinct as the embryo grows and definite tissues, organs and systems develop. However, even in the most complex of triploblasts, e.g. mammals, all the tissues of the body can be described as being ectodermal, endodermal or mesodermal in origin. Triploblasts are bilaterally symmetrical (i.e. have dorsal, ventral and lateral surfaces and usually have a recognisable anterior. i.e. head, end).
This type of organisation is a prerequisite for efficient locomotion.

Flatworms (Platyhelminthes) are the simplest triploblasts, and the mesoderm does not become highly elaborated (in fact much of it remains as 'packing tissue' of unspecialised cells) but true muscle tissue and the reproductive system are both derived from it. As there is no blood system the size and shape of the body must be such that food and oxygen can reach all the living cells merely by diffusion. Thus the body is always flattened and the gut, when present, is always very branched and ramifying (it has only one opening, as in Coelenterates). Almost all Platyhelminthes are hermaphrodite, although cross-fertilization usually takes place. The reproductive system is always very complicated.

As in Coelenterates, the endoderm forms the lining of the gut and is digestive and absorptive in function. The endodermal cells can engulf food particles by means of pseudopodia.

The ectoderm again provides the outer protective layer of the body, and gives rise to the neuro-sensory system. But in invertebrate Triploblasts it also gives rise to definite excretory organs, *nephridia*. In the Platyhelminthes these take the form of branched ectodermal tubules, terminating in a system of 'flame cells' at one end and opening to the exterior at the other. They are believed to be mainly concerned with osmo-regulation, i.e. they eliminate excess water from the body, but it is probable that the fluid contains excretory products, including nitrogenous waste.

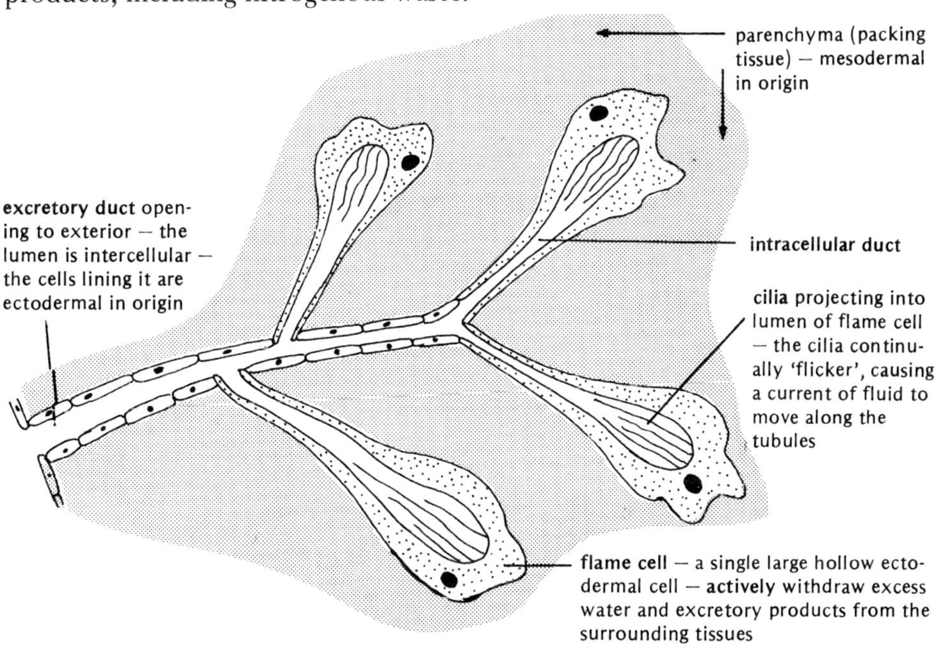

Diagram to show the basic principles of the nephridial system of Platyhelminthes

Phylum Platyhelminthes

A central nervous system is present in all triploblasts. In the Platyhelminthes it consists of paired cerebral ganglia and two or more solid non-ganglionated nerve cords. Sensory and motor nerve fibres constitute the peripheral nervous system connecting receptors and effectors to the CNS, and impulses therefore follow definite pathways. Due to the complex system of synapses within the CNS, a much more elaborate interconnecting system is possible than in Coelenterates. Responses to a localised stimulus can be manifest in any part of the body, and more flexibility in response is can react as a whole to a single stimulus, and more flexibility in response is possible. In Platyhelminthes the cerebral ganglia are concerned solely with receiving sensory impulses from the sense organs in the head and initiating appropriate motor responses. They are not equivalent to a 'brain', since the term 'brain' implies an additional role as an integration centre for co-ordinating movement and controlling complex behavioural reactions. An extensive system of nerve nets is still present in Platyhelminthes, and many of the activities of the body are controlled by these and do not involve the CNS.

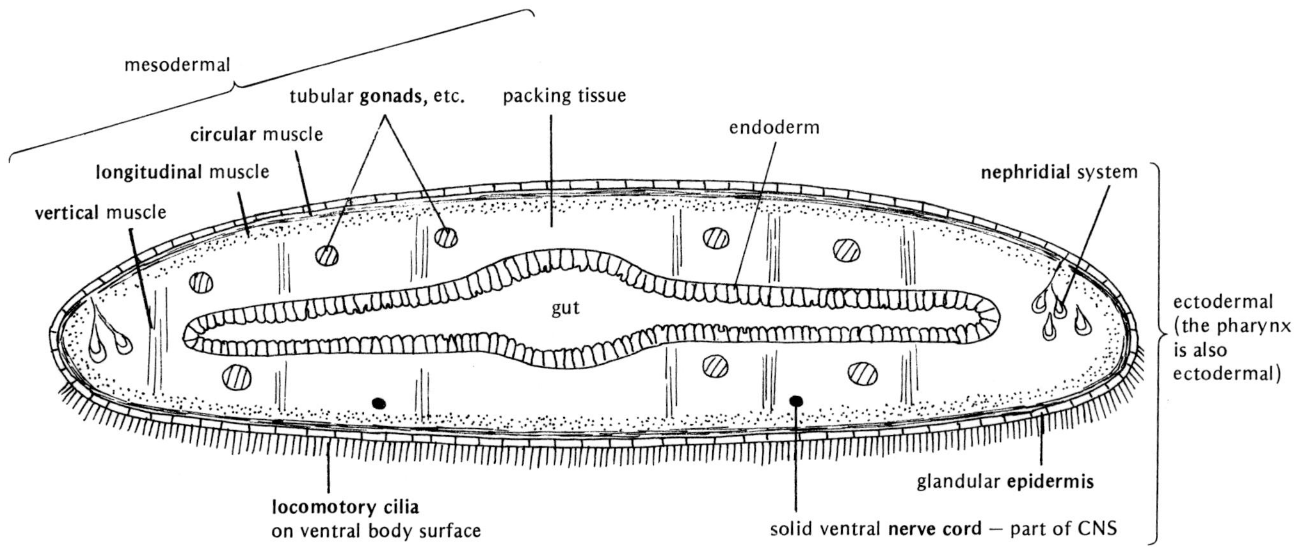

TS generalised Platyhelminth

General characteristics

Aquatic – free-living or parasitic.

Triploblastic – bilaterally symmetrical – body flattened dorso-ventrally.

Gut (when present) usually very branched. Only one opening to exterior.

CNS present (= paired cerebral ganglia and 2 or more ventral solid non-ganglionated nerve cords).

Nephridial system of flame cells linked by branching ducts which open to the exterior by pores.

No blood system.

Hermaphrodite. Very complex reproductive system.

Class Turbellaria	Class Trematoda (flukes)	Class Cestoda (tapeworms)
Free living, aquatic	Parasitic – life cycle involving a vertebrate and invertebrate hosts	Endoparasitic in alimentary canal – life cycle usually involving two vertebrate hosts
No cuticle; epidermis present	Thick cuticle replaces epidermis	Thick cuticle replaces epidermis
Epidermis ciliated	No cilia, but spines often present	No cilia or spines
Active locomotion	Feeble locomotion. Sense organs reduced.	Very feeble movement. Sense organs reduced.
Gut present – usually much branched, pharynx often eversible	Much branched gut	No gut
Leaf-like body	Leaf-like body	Body usually strobilated – i.e. proglottides present, and growth continues throughout life
No suckers	2 suckers – oral and ventral	Many suckers and hooks on scolex (head)
Development usually direct	Complex larval stages	Complex larval stages
E.g. the following genera:		
Planaria	**Fasciola** (liver fluke)	**Taenia** (tapeworm)

CLASS TURBELLARIA

e.g. *Planaria sp.*

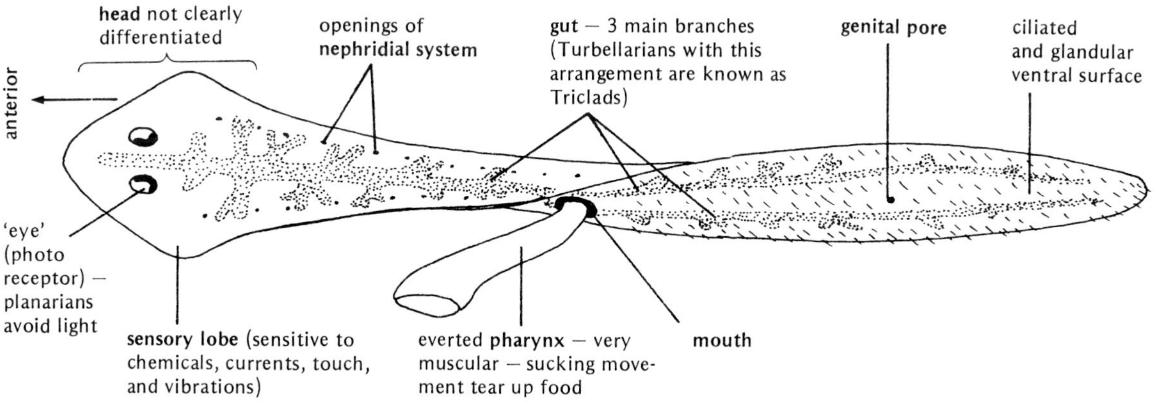

External features (Enlarged. May reach up to 1 cm in length)

Notes

1. Very common in fresh water ponds, ditches, etc.
2. Glides over surfaces by ciliary action – secretes mucus on to substratum as it moves. During rapid locomotion the longitudinal muscles on each side of the body contract alternately, causing a series of ripples to pass down the body, driving the animal forwards. Contractions tend to pull the head from side to side, causing the planarian to follow a very irregular path; this helps to keep it hidden under leaves, etc. It cannot swim.
3. Carnivorous – actively seeks food (small living prey, as well as dead bodies of larger animals) – mounts the food, presses it against substratum, entangles it in mucus, and sucks up microscopic fragments by means of muscular eversible pharynx.
   Digestion mainly intracellular – the particles ingested are small enough to be engulfed by the cells lining the gut.
4. Has considerable powers of regeneration.
5. The cup-shaped pigment cells of the eyes screen the light sensitive cells from all sides but one, so that the direction as well as the intensity of light can be appreciated.

Phylum Platyhelminthes
Class Trematoda
Genus *Fasciola*

CLASS TREMATODA

e.g. *Fasciola hepatica* (liver fluke of sheep)

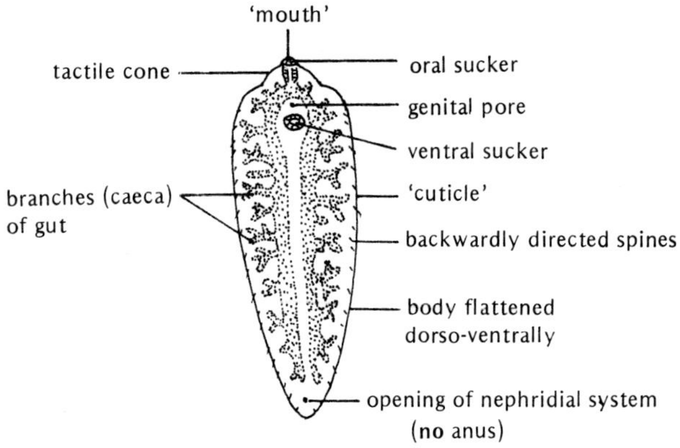

Adult ventral view (x2)
(Body leaf-like, 1 – 2½ cm in length)

Notes

1. The adult fluke is an internal parasite found curled up in the bile duct system of sheep, and responsible for 'liver rot'.
2. It feeds by sucking fluids, particularly blood, from the walls of the smaller bile ductules. Thus its food consists mainly of blood from the hepatic portal system of the host, blood which is extremely rich in already digested food.
3. Limited powers of locomotion – can 'loop' along by means of its suckers.
4. The cuticle is thick but permeable and allows for gaseous exchange, etc. It is mesodermal in origin, and is living. It protects the fluke from the digestive juices, antibodies, etc. of its host. (The epidermis is lost during development.)

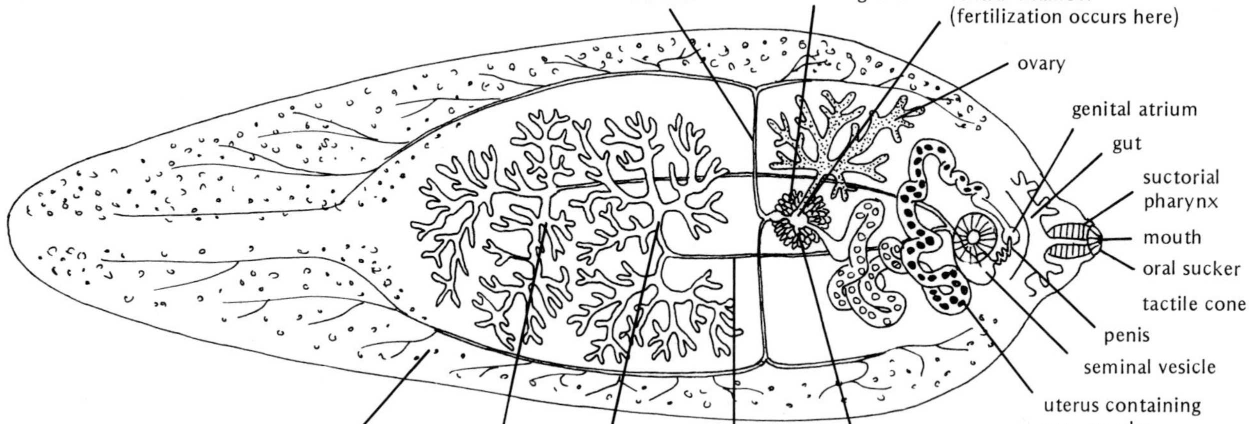

Reproductive system

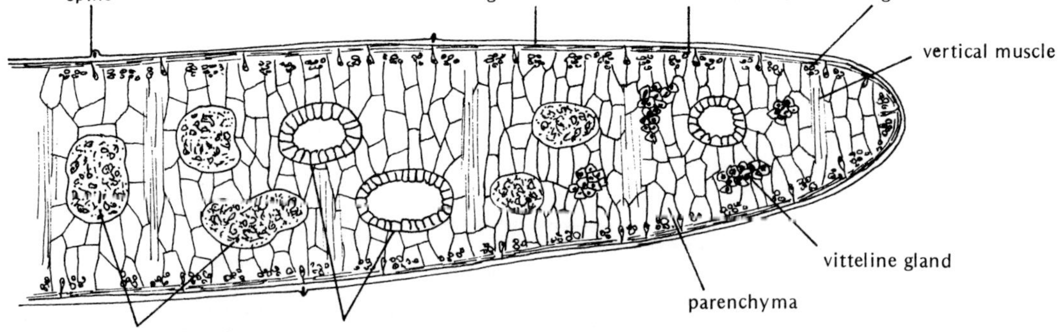

TS in region of testes

Phylum Platyhelminthes

Class Trematoda

Genus *Fasciola*

Life cycle of Fasciola hepatica

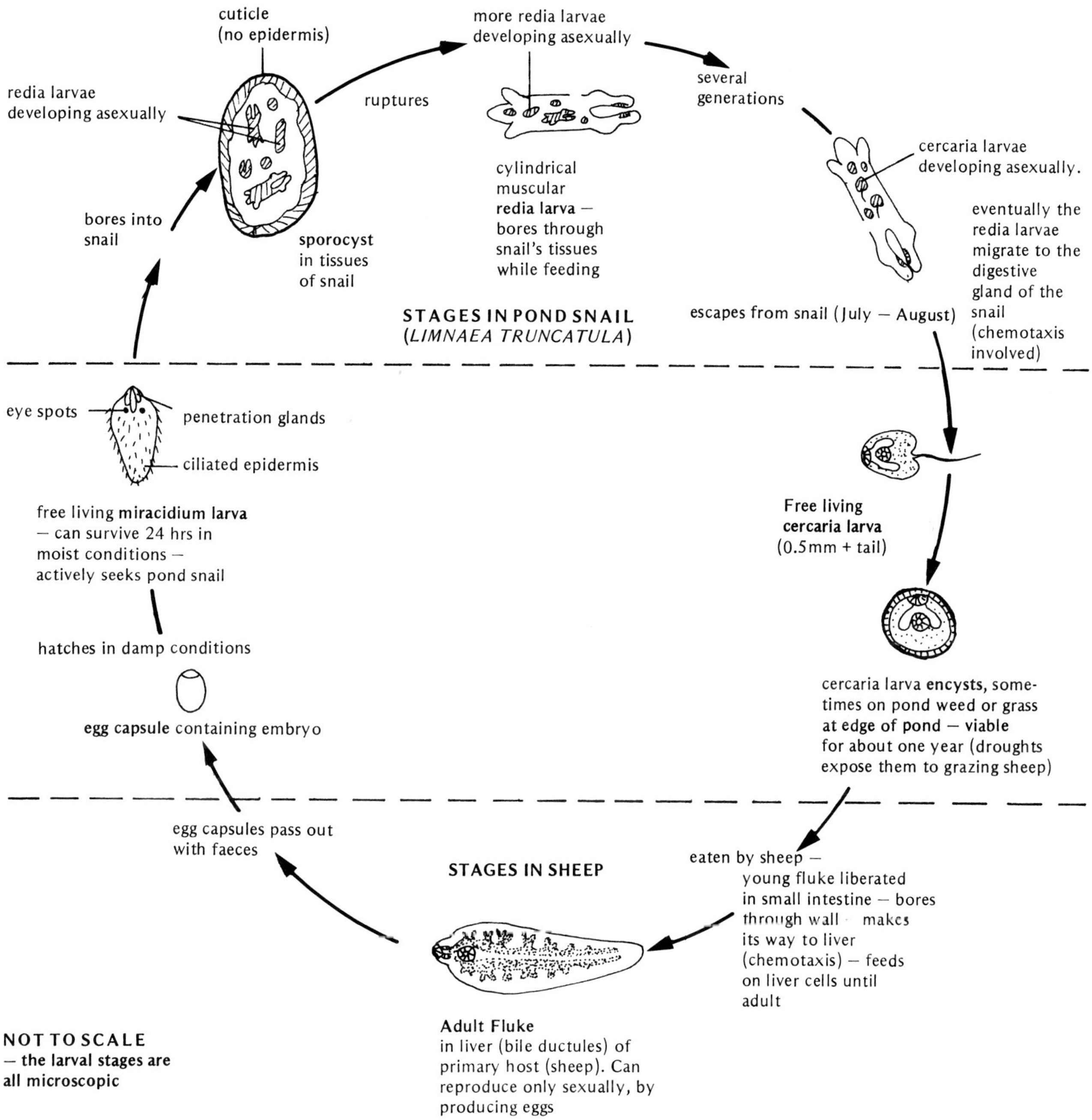

Phylum Platyhelminthes

Class Trematoda

Genus *Fasciola*

Adaptations of structure and life history to a parasitic mode of life

Parasitism is a secondary state; adaptations shown by parasites are modifications or loss of structures characteristic of the free living members of the phylum. The modifications are most marked in internal parasites. The adult Fasciola shows the following structural adaptations:

i. The possession of adhesive devices for attachment to host (suckers and spines).
ii. Reduction of trophic (feeding) organs – no elaborate eversible pharynx.
iii. Reduction of locomotory power – no cilia. (The fluke is surrounded by continuously available food.)
iv. Reduction of neuro-sensory system, since there is very little fluctuation in the conditions to which the fluke is subjected.
v. Possession of a cuticle resistant to the defence reactions of the host.
vi. Considerably increased reproductive powers; the abundance of food allows for the production of very many eggs.

Although an adult fluke leads a life free from most of the usual hazards confronting animals, the problems involved in the dispersal of offspring are much greater, and inevitably the establishment of offspring in new hosts involves considerable losses. Therefore:

i. An enormous number of eggs is produced.
ii. A vector (secondary host) serves to convey the larval stage from one primary host to another.
iii. An asexual reproductive phase occurs in the secondary host, and ensures that a single successful larva eventually gives rise to thousands of further larvae, each capable of infecting a sheep.

Remaining adaptations essential for the successful completion of the life cycle are as follows:

iv. The eggs are enclosed in protective capsules so that they are not digested before leaving the primary host.
v. The larvae which hatch from the capsules can actively seek out and penetrate a suitable secondary host.
vi. The many larvae which leave the snail can encyst, and thus remain viable for a considerable time.
vii. The cyst protects the young fluke from the acid gastric juices of the sheep, so that it is not liberated until it has reached the small intestine.

Note

The parasite is physiologically adapted at all stages of its life cycle to survive the diverse situations it encounters during its transference through these totally different environments.

CLASS CESTODA

e.g. *Taenia solium* (human pork tapeworm)

Phylum Platyhelminthes
Class Cestoda
Genus *Taenia*

External features (Approx. life size)

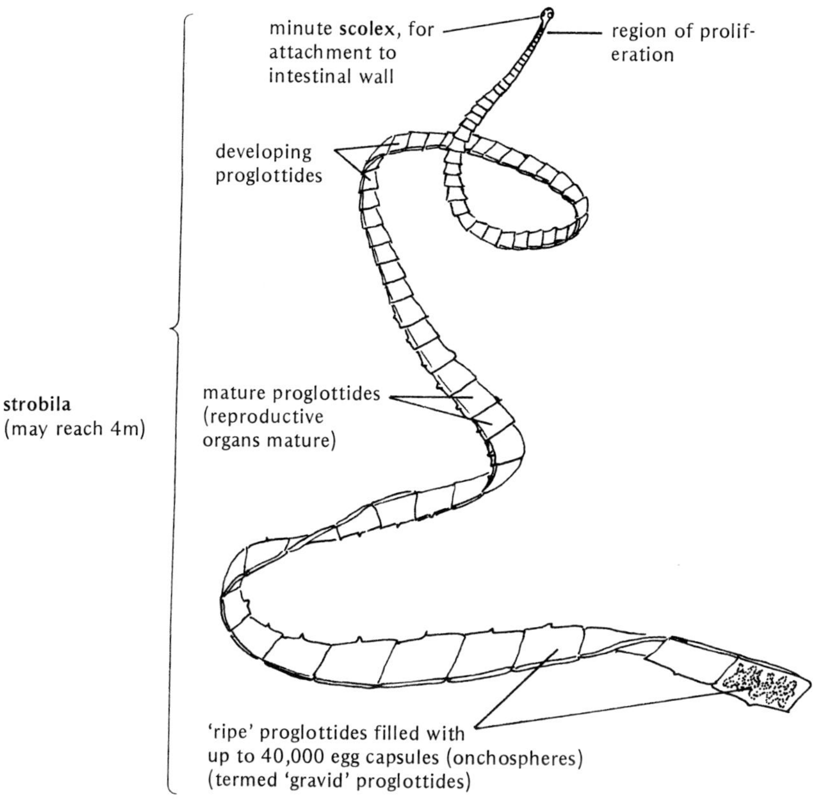

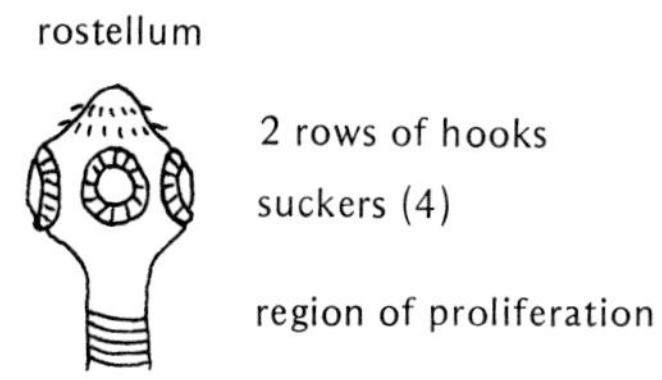

Detail of scolex (Enlarged)

The tapeworm may cause emaciation of the host, and the toxins it produces can cause anaemia, nausea, and nervous disorders.

As long as the scolex remains attached, it will continue to produce new proglottides – the scolex must be destroyed to eliminate the tapeworm permanently.

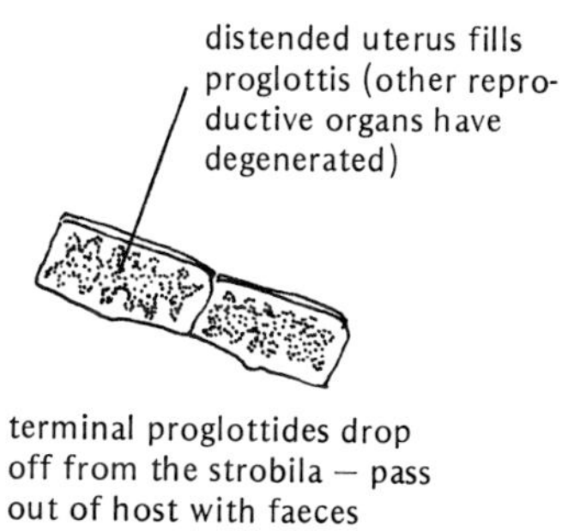

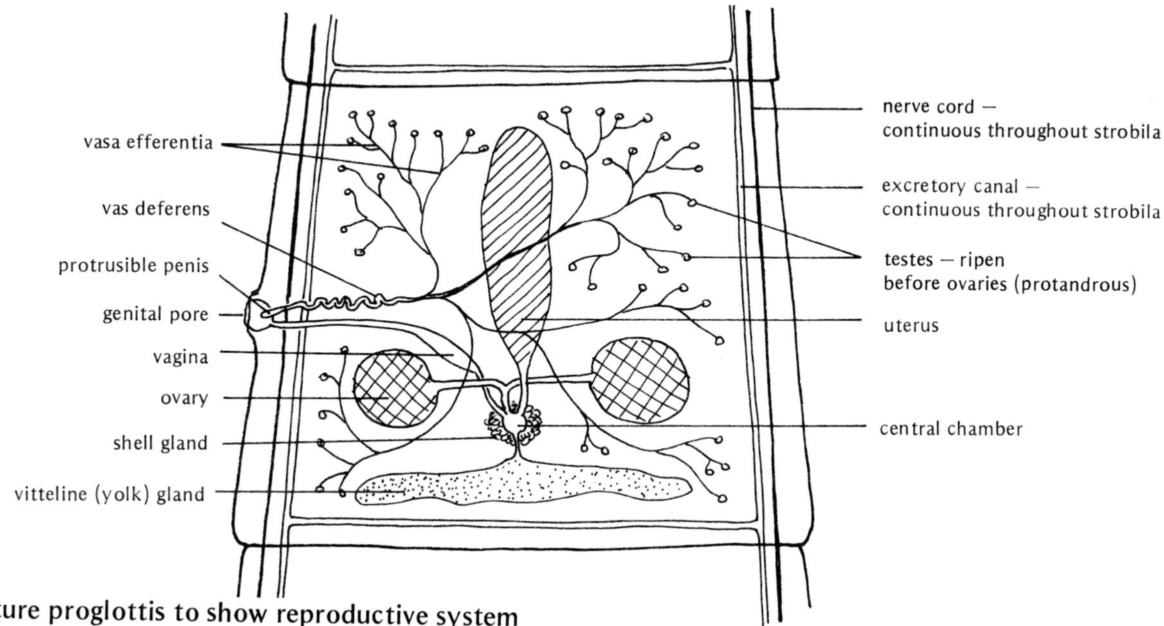

Mature proglottis to show reproductive system
(Highly diagrammatic)

Phylum Platyhelminthes

Class Cestoda

Genus *Taenia*

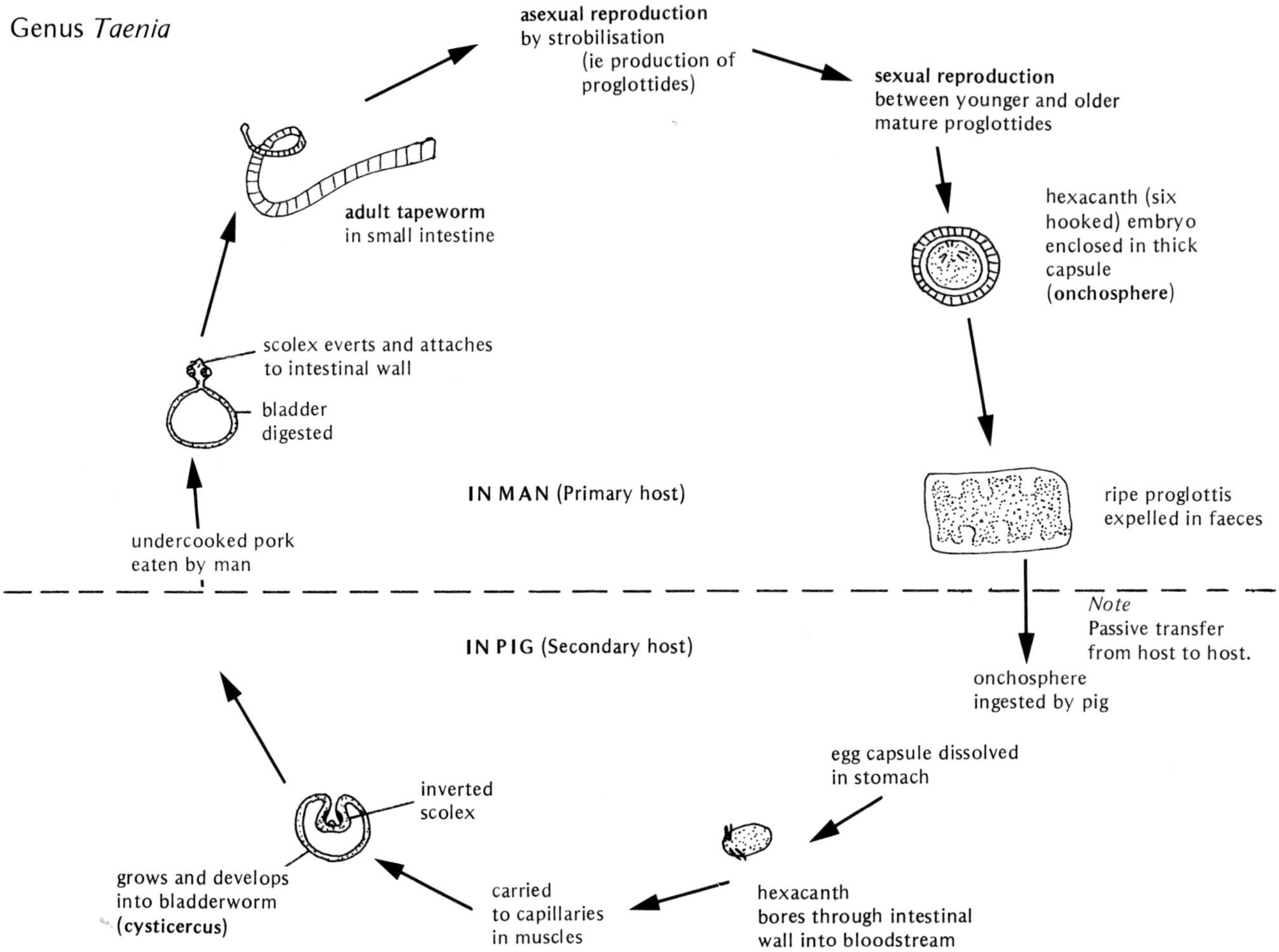

Taenia solium
Life cycle (Not to scale)

Notes on adult tapeworm

1. No mouth, no gut. Digested food is absorbed from the surroundings via the 'cuticle', the surface of which is increased by microvilli.
2. No sense organs.
3. Thick, living 'cuticle' resistant to the digestive juices of host.
4. Respiration anaerobic (surroundings have very low oxygen tension) – lactic acid excreted via cuticle.
5. Although self-fertilization of an individual proglottis is possible, the male organs develop before the female and usually sperm are transferred from one section of the worm to the proglottides further along the strobila, while it is folded back on itself. Cross fertilization between two tapeworms is unlikely, since a single host cannot usually support more than one large tapeworm.
6. Each proglottis is biologically complete in itself and has little functional dependence on another. Each is basically similar in structure to an individual flatworm. Although the nerve cords and excretory canals are continuous throughout the length of the worm, the strobila does not, in most respects, function as an integrated whole. It can be of any length, and the 'segments' are of different ages, sizes and degrees of development.

10. Phylum Nematoda (roundworms)

General characteristics

Ubiquitous – occur in *enormous* numbers in a great variety of habitats.

Free-living forms are usually minute – occur in soil, at the bottom of lakes, rivers, seas, etc. and in all rotting materials.

Parasitic forms are usually larger; they occur as internal parasites in practically all plants and animals.

Triploblastic, non-segmented, acoelomate, but body cavity present due to breakdown of 'parenchyma'.

Uniformly cylindrical elongated body, tapering at both ends.

Thick, slightly flexible, tough cuticle.

No circular muscles in body wall – 4 bands of longitudinal muscle. Body can bend only in the dorso-ventral plane. Locomotion by thrashing about.

No cilia.

No blood system.

Unbranched gut with mouth and anus – suctorial pharynx – microphagus or fluid feeders (cannot ingest large particles of food).

CNS present, but nervous system poorly developed.

No well developed sense organs, merely sensory papillae around mouth – sometimes innervated bristles present.

Sexes separate (i.e. dioecious) – gonads tubular.

Development direct. Cuticle shed 4 times during growth to adult size.

The cuticle of a nematode worm is rather like a straight-jacket, and prevents the worm from accumulating anything in its body; the worm cannot gorge itself with occasional large meals. It has to spend all its time feeding and is therefore not free to employ itself in many other activities. Similarly the female has to lay its eggs continuously.

e.g.
Ascaris lumbricoides (in human intestine)
Proleptus sp. (in gut of dogfish)
Filaria bancrofti (in the lymphatic system of humans, causing elephantiasis)
Rhabditis maupasi (juvenile stage in the nephridia of earthworms)
also *hook worms* and the *guinea worm* in humans and *eelworms* of crops.

Ascaris lumbricoides (human roundworm)

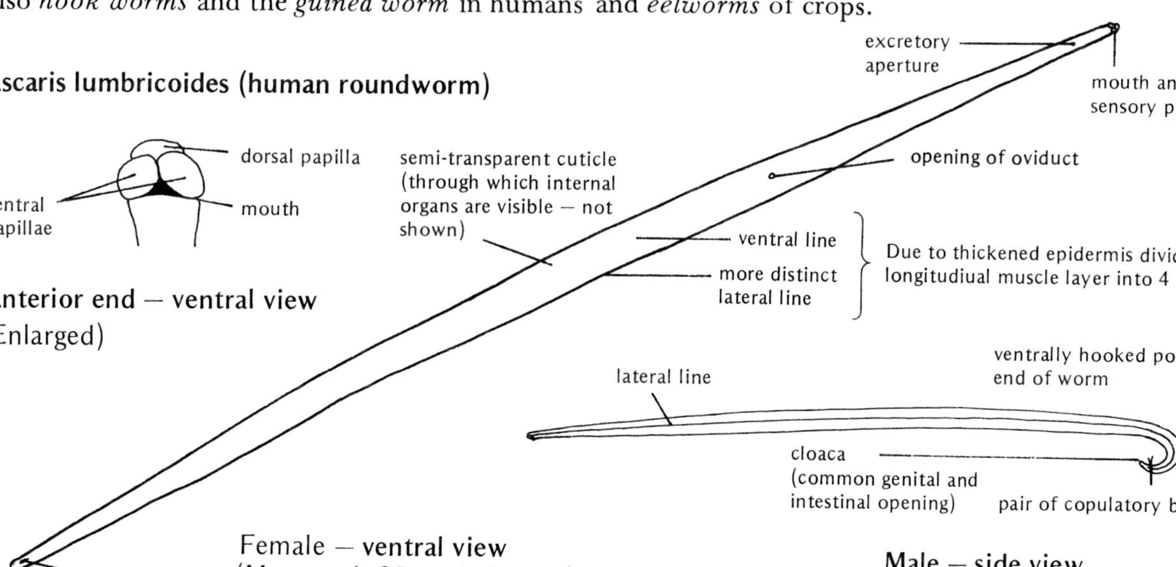

Anterior end – ventral view
(Enlarged)

Female – **ventral view**
(May reach 25 cm in length)

Male – side view
(May reach 15 cm in length)

Life history (no vector involved)

More than 1000 eggs are laid daily by the female. They pass out of the host with the faeces. The young worm is ready to infect a new host after moulting once within the egg-shell. Having been swallowed by a human host, it hatches and bores its way through the lining of the small intestine into the blood stream. Growing considerably, it is eventually carried to the lungs, where it penetrates into a bronchial tube. It remains here for about a week, moulting twice, and then migrates up the trachea to the mouth, to be swallowed back into the stomach. Once in the small intestine it grows for a further month, and becomes sexually mature at its final moult.

11. Phylum Rotifera (wheel animalcules)

General characteristics

Very common in fresh water. In fact they are the most common freshwater Metazoans.

Microscopic, about the same size as many Protozoa.

Triploblastic, acoelomate, non segmented.

Mostly free-living – few parasitic.

Shape very variable, ranging from worm-like forms to flower-like forms.

Usually divided into head, trunk and tapering adhesive post-anal foot (containing cement glands).

Crown of cilia at anterior end, concerned in locomotion (including swimming) and creating food current.

Body enclosed in flexible cuticle, sectioned to allow for telescoping during contraction of body.

Foot provides anchorage during feeding. Food consists of algae, bacteria, small protozoa, etc.

Alimentary canal with mouth and anus – grinding pharynx, bearing teeth.

Simple eyes often present.

Nephridial system with flame cells – bladder also present.

No blood system. Microscopic size precludes the necessity of a transport system; diffusion suffices.

Body composed of constant number of cells. These lack cell membranes, i.e. the body is a *syncytium*. Cell constancy is so marked that each organ is composed of a certain unvarying number of cells, each nucleus occupying a definite position. Regeneration does not occur.

Sexes separate. Male individuals small, degenerate and short lived, produced only during sexual season. Parthenogenesis occurs for remainder of year; females give rise to other females by means of unfertilized eggs.

Both adults and sexually produced eggs are extremely resistant to desiccation, thus rotifers can be found in situations which are only occasionally wet.

e.g. *Philodina acutrionus*

(Approx. 0·5 mm)

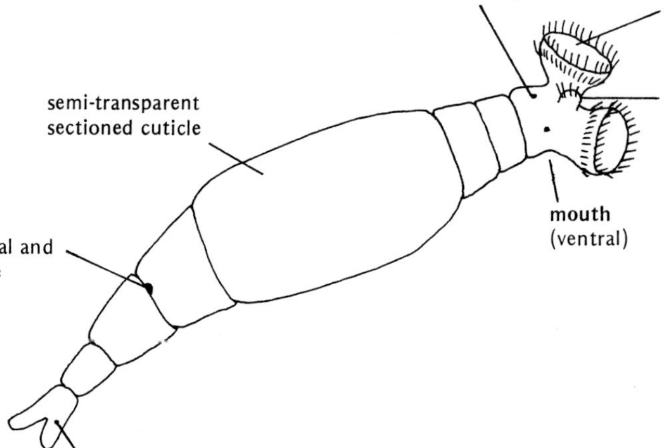

12. Coelomate Animals

Coelomates are triploblastic animals in which the mesoderm is split into two layers separated by a fluid-filled space (the coelom). The inner layer of mesoderm surrounding the gut is known as splanchnic mesoderm. The outer layer is known as somatic mesoderm.

The gut and its associated muscles are thus separated from the body wall. As a result:

i. Locomotion and digestive movements can take place simultaneously and independently. Parts of the alimentary canal can become specialised, and the muscles of the body wall can be grouped in association with locomotory organs. Mutual displacement of organs can occur.

ii. The presence of the coelom and the increased muscular differentiation means that the body cannot be very flattened. Direct diffusion of food and gases throughout the body cannot therefore take place at an adequate rate; a conducting system (usually a blood system) must be present.

iii. The presence of a blood system makes it possible for certain parts of the body to be set aside as specialised respiratory surfaces and so allows for greatly increased structural complexity and the attainment of a much larger size.

In all coelomates the gut has a mouth and an anus, and most coelomates are metamerically segmented.

TS body of a primitive Coelomate (Very diagrammatic)

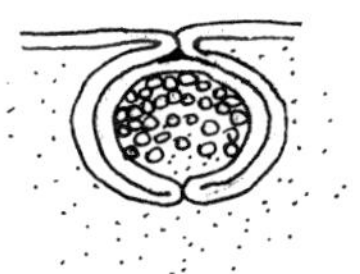

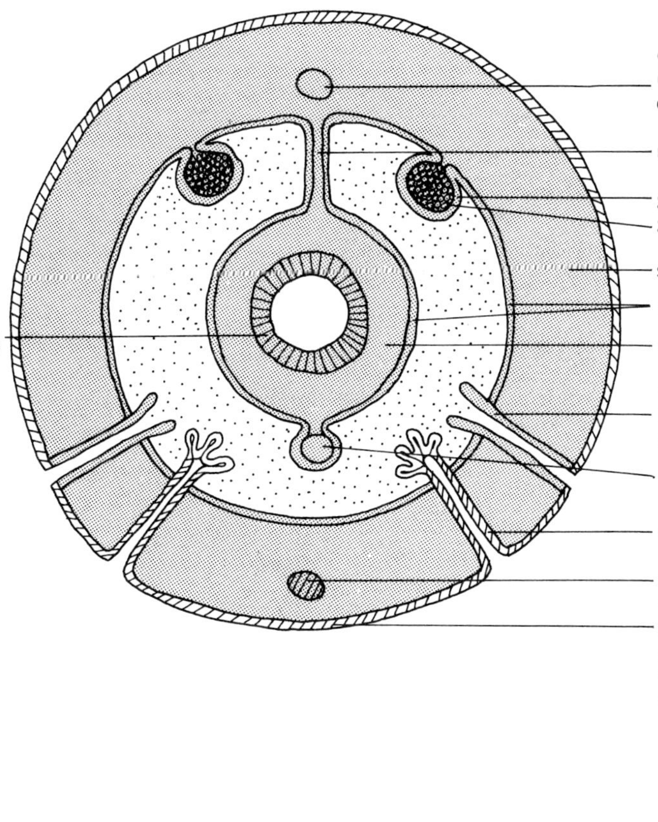

Metameric segmentation

The body of a metamerically segmented animal is made up of a series of similar segments, the number of which is constant. The segmentation occurs in the developing embryo and ideally, once development is completed, no further segments are added to the body, although the animal continues to grow until it reaches adult size. Thus all the segments of the body are the same age.

Again ideally, all the segments are structurally identical so that the various organs are repeated throughout the body, i.e. each segment contains a similar pattern of blood vessels, nerves, excretory organs, gonads, etc. However, a segment is *not* a self-contained unit and all the segmental structures function inter-dependently. The activities of each segment are co-ordinated with all the others, so that, unlike a proglottis, every segment forms an integral part of the complete body, which functions as a whole.

Metamerism is always obscured by modification of certain regions of the body, particularly at the anterior end (due to cephalisation – the development of the head) and by the development of limbs, but it is practically complete in some animals, e.g. some Annelida. In others, e.g. Vertebrata, the basic segmentation of the body becomes almost completely obscured during development, and is noticeable only in the embryo.

This fundamental regularity and serial repetition of structures lends itself more readily to the modification and adaptation of various parts of the body than does an organisation where there is no serial repetition (as in Mollusca and Echinodermata), i.e. it opens the way for wide variations in form and structure.

L.S. Metamerically segmented animal

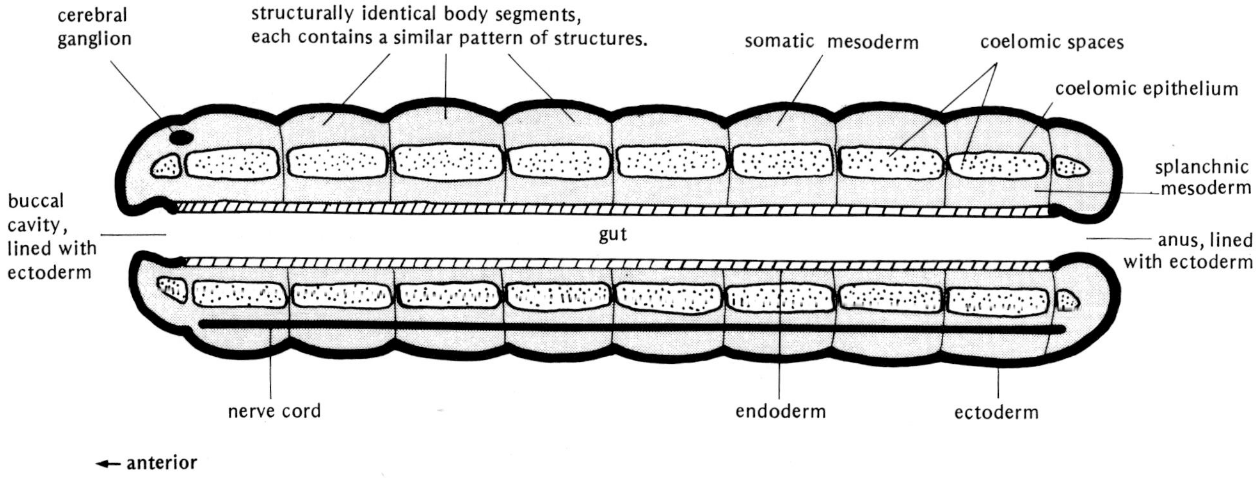

13. Phylum Annelida

General characterisitcs

Segmented worms – soft bodied.

Typically aquatic.

Coelomate – coelom forms the body cavity.

Clearly metamerically segmented.

Thin cuticle, composed of collagen – not waterproof – gaseous exchange takes place over entire body surface.

Chaetae (bristles) typically present – composed of chitin.

CNS consists of solid ventral double ganglionated nerve cord and paired dorsal cerebral ganglia.

Blood system present – closed (i.e. blood confined to blood vessels).

Excretory system provided by segmentally arranged paired nephridia.

Long tentacles, segmental parapodia, or external gills may be present – these increase the respiratory surface.

Simple eyes typically present.

Class: Polychaeta (bristleworms)	**Class: Oligochaeta (earthworms)**	**Class: Hirudinea (leeches)**
Marine	Damp terrestrial/fresh water	Fresh water/damp terrestrial /marine. Ectoparasitic
External segments correspond with inner septa	External segments correspond with inner septa	External 'segments' more numerous than inner septa
Distinct head with appendages	Indistinct head	Indistinct head
Eyes usually present	No eyes	Eyes present
Parapodia usually present, bearing many chaetae	No parapodia, few small chaetae	No parapodia, no chaetae
No suckers	No suckers	2 suckers, anterior and posterior
Coelom well developed	Coelom well developed	Coelom occluded – body almost solid
Sexes separate	Hermaphrodite (but cross fertilization)	Hermaphrodite (but cross fertilization)
Gonads extend throughout length of body	Gonads confined to few genital segments	Gonads confined to few gential segments
No clitellum	Clitellum present – cocoons produced	Clitellum present – cocoons produced (clitellum may not be obvious)
External fertilization	Fertilization in cocoon	Fertilization in cocoon
Larval stage present (trochophore)	No larval stage (direct development)	No larval stage (direct development)
Locomotion by lateral undulations of body	Locomotion by waves of thinning and extension alternating with waves of expansion	Swimming by vertical undulations; locomotion on land by looping
E.g. the following genera: **Nereis** (ragworm) **Sabella** (tubeworm) **Arenicola** (lugworm)	**Lumbricus** **Allolobophora** (earthworms)	**Hirudo** (leech)

Phylum Annelida
Class Polychaeta
Genus *Nereis*

CLASS POLYCHAETA

e.g. *Nereis diversicolor*

Yellow/brown in colour with green line along each side and red blood vessel in mid-dorsal, hence the name *diversicolor.*

External features – dorsal view x4. 7-10 cm long

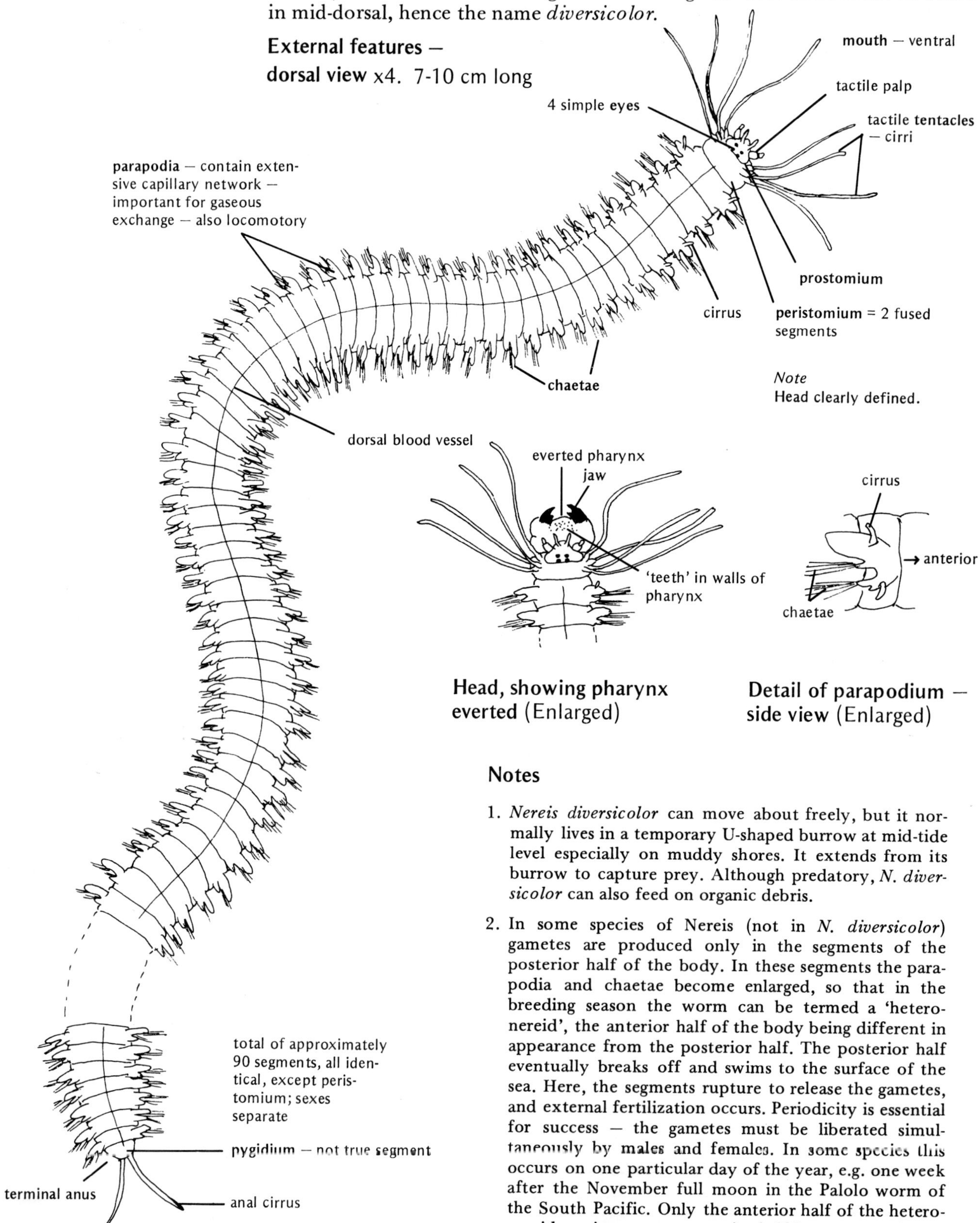

Head, showing pharynx everted (Enlarged)

Detail of parapodium – side view (Enlarged)

Notes

1. *Nereis diversicolor* can move about freely, but it normally lives in a temporary U-shaped burrow at mid-tide level especially on muddy shores. It extends from its burrow to capture prey. Although predatory, *N. diversicolor* can also feed on organic debris.
2. In some species of Nereis (not in *N. diversicolor*) gametes are produced only in the segments of the posterior half of the body. In these segments the parapodia and chaetae become enlarged, so that in the breeding season the worm can be termed a 'heteronereid', the anterior half of the body being different in appearance from the posterior half. The posterior half eventually breaks off and swims to the surface of the sea. Here, the segments rupture to release the gametes, and external fertilization occurs. Periodicity is essential for success – the gametes must be liberated simultaneously by males and females. In some species this occurs on one particular day of the year, e.g. one week after the November full moon in the Palolo worm of the South Pacific. Only the anterior half of the heteronereid survives; a new posterior half is regenerated.

CLASS OLIGOCHAETA

e.g. *Lumbricus terrestris* (earthworm)

Phylum Annelida
Class Oligochaeta
Genus *Lumbricus*

External features

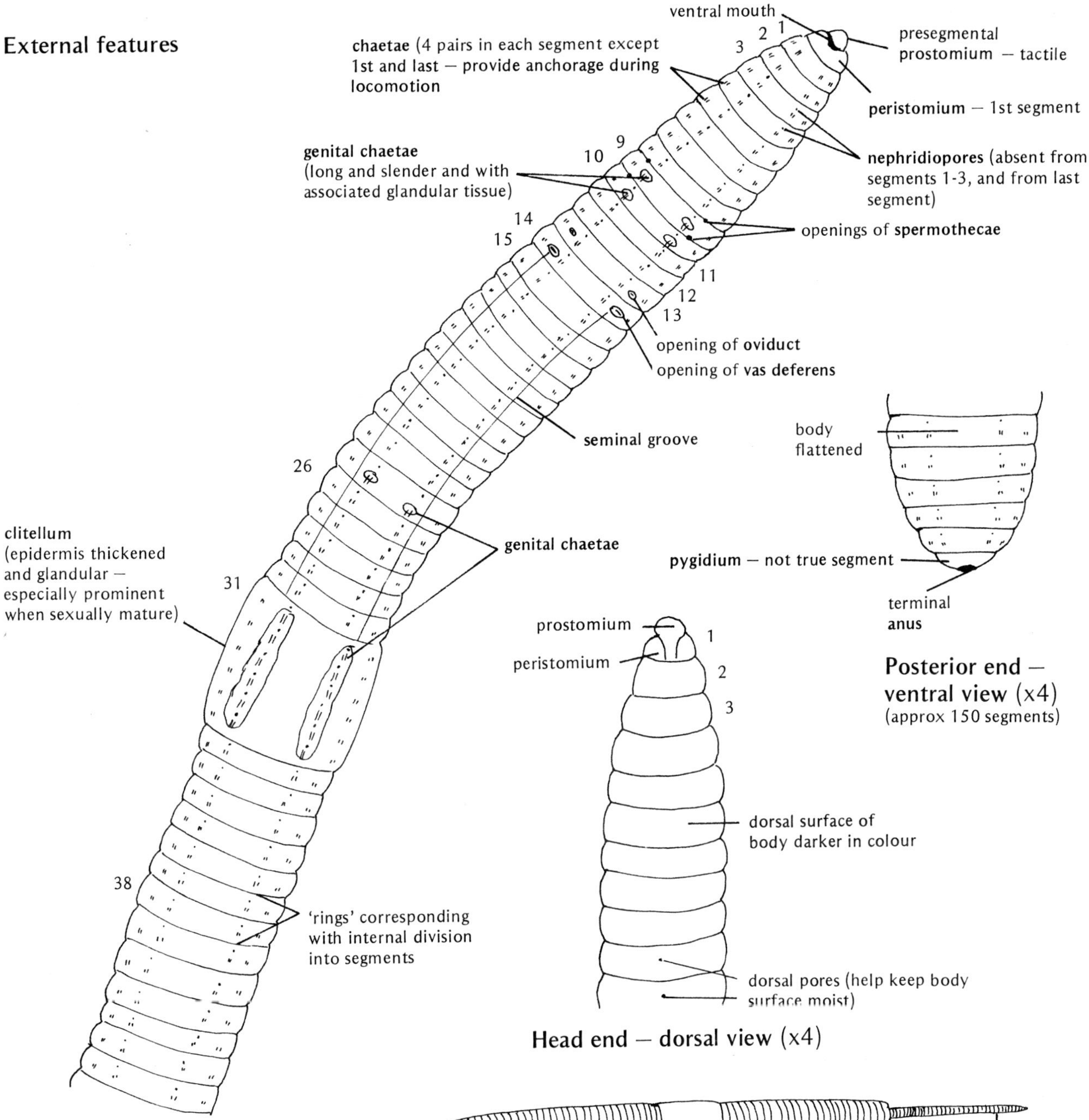

Posterior end – ventral view (x4)
(approx 150 segments)

Head end – dorsal view (x4)

Anterior end – ventral view (x4)

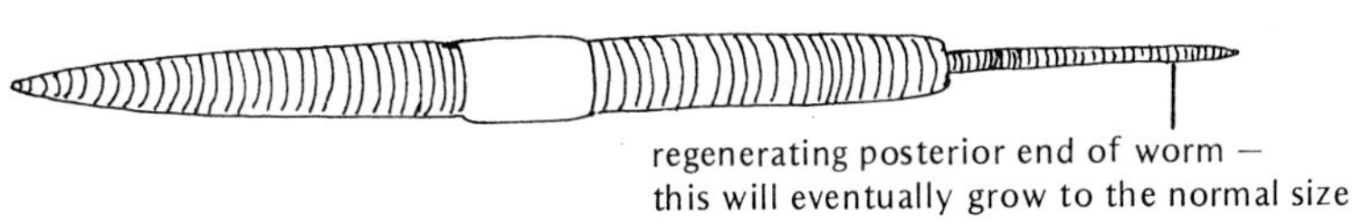

Diagram to show regeneration of posterior half (x1)

Phylum Annelida
Class Hirudinea
Genus *Hirodo*

CLASS HIRUDINEA

e.g. *Hirudo medicinalis* (medicinal leech)

External features – ventral view (x2)

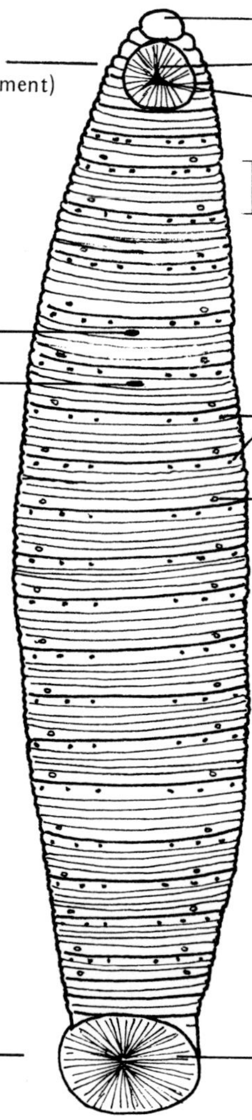

Notes

1. Found in marshes, ponds and streams, but not common in Britain.
2. Swims well by *vertical* undulations of the body, which is slightly flattened dorso-ventrally.
3. Can also move by looping (using suckers).
4. Feeds on blood, sucked mainly from mammals. Its salivary glands secrete 'hirudin' which prevents blood coagulating during feeding. One meal can nourish the leech for many months, the digestive tract having large side pouches for storing food.

14. Phylum Arthropoda

General characteristics

Extremely widespread; largest and most successful animal phylum.

Free living, parasitic and sedentary forms occur.

Bilaterally symmetrical, metamerically segmented.

Coelomate, but body cavity consists of a *haemocoel*, and the coelom is reduced to the cavities of the gonads and the excretory organs. (This is because during embryonic development the blastocoel remains extensive, whereas in most other coelomates it persists only as the main dorsal and ventral blood vessels.)

Open blood system – *dorsally* situated heart present.

Cephalisation pronounced. Body can be differentiated into *head* of 6 segments, *thorax*, and *abdomen*.

Chitinous jointed exoskeleton present. This cannot grow, therefore *ecdysis* occurs during development, i.e. moulting.

Basically one pair of jointed appendages per body segment, but reductions occur, some of those of the head segments being modified to form mouth parts (i.e. 1 pair of mandibles, and 2 pairs of maxillae).

No nephridia; coelomoducts provide the excretory organs.

Muscles of body wall occurs in separate bundles, not as continuous layers.

Anus terminal. Alimentary canal lined with cuticle to a varying extent (due to involutions of the exoskeleton at each end).

CNS consists of solid double ventral segmentally ganglionated nerve cord, together with well developed dorsal cerebral ganglia.

Compound eyes may be present; compound eyes are unique to Arthropods. Tactile setae (bristles) typically present.

No cilia.

There are five main classes: *Crustacea, Insecta, Arachnida, Diplopoda* (millipedes) and *Chilopoda* (centipedes).

CLASS CRUSTACEA

General characteristics

Typically aquatic – gills present.

Exoskeleton usually strong, massive – hardened by lime salts (calcite).

Head and thorax fused to form *cephalothorax*, which is often protected by a *carapace*.

Head bears 2 *pairs* antennae in front of mouth, and 3 pairs mouth parts.

Variable number of thoracic segments, each bearing 1 pair of appendages.

Variable number of abdominal segments, each bearing 1 pair of appendages.

Thoracic and abdominal appendages modified for *various* purposes, e.g. food capture, swimming, walking, respiration, reproduction, etc. They are primitively biramous. (see over.)

Excretion by antennal and maxillary glands (derived from coelomoducts). Also via gills and gut wall.

Respiratory pigment (haemocyanin or occasionally haemoglobin) may be present, dissolved in plasma.

Sexes usually separate.

Nauplius larval stage typically present.

There are three main sub-classes, but several other minor sub-classes exist, including the Copepoda, of which the genus Cyclops is a representative.

Phylum Arthropoda

Class Crustacea

Sub-class Branchiopoda	Sub-class Cirripedia	Sub-class Malacostraca
Fresh water	Marine	Usually marine
Compound eyes present – may be fused to form single median eye	No eyes	1 pair compound eyes – usually stalked
Carapace of 2 valves – may enclose entire body	Carapace of calcareous plates	Carapace, when present, not covering abdomen
At least 4 pairs of uniform thoracic appendages for filter feeding	6 pairs of thoracic appendages, for filter feeding	8 pairs of thoracic appendages for walking
Abdomen bears no appendages – may be very reduced	No abdomen	Abdomen of 7 segments, bearing appendages for swimming
Sexes separate – parthenogenesis common	Hermaphrodite, but cross-fertilization	Sexes separate
Development either direct or with nauplius larva	Larval stages present (nauplius and cypris)	Development usually direct
Diverse group, exhibiting much variety of form	Sedentary, fixed	Diverse group, but common basic body plan
E.g. the following genera: **Daphnia** } (water **Simocephalus** } fleas) **Chirocephalus** (fairy shrimp)	 **Balanus** (barnacle) **Sacculina** (a parasite on crabs)	 *Order Decapoda:* **Astacus** (crayfish) **Carcinus** (crab) *Order Isopoda:* **Oniscus** (woodlouse) *Order Amphipoda:* **Gammarus** (freshwater shrimp)

All the amazingly complicated forms of crustacean limb and mouth parts have been derived from a basic biramous form of appendage:

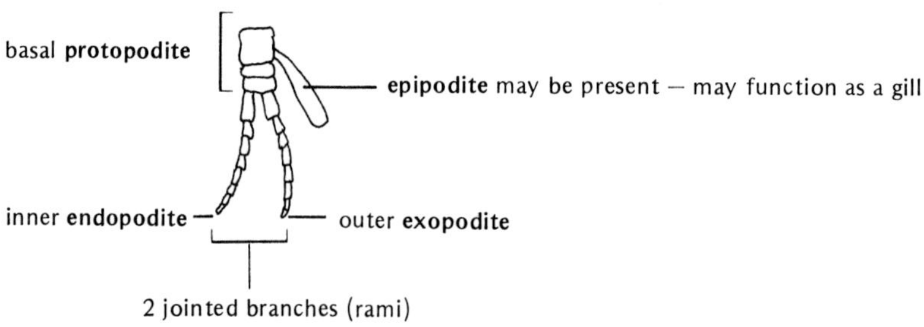

Generalised crustacean appendage (Very diagrammatic)

SUB-CLASS BRANCHIOPODA

Phylum Arthropoda
Class Crustacea
Sub class Branchiopoda
Genus *Daphnia*

e.g. *Daphnia sp.* *(water flea)*

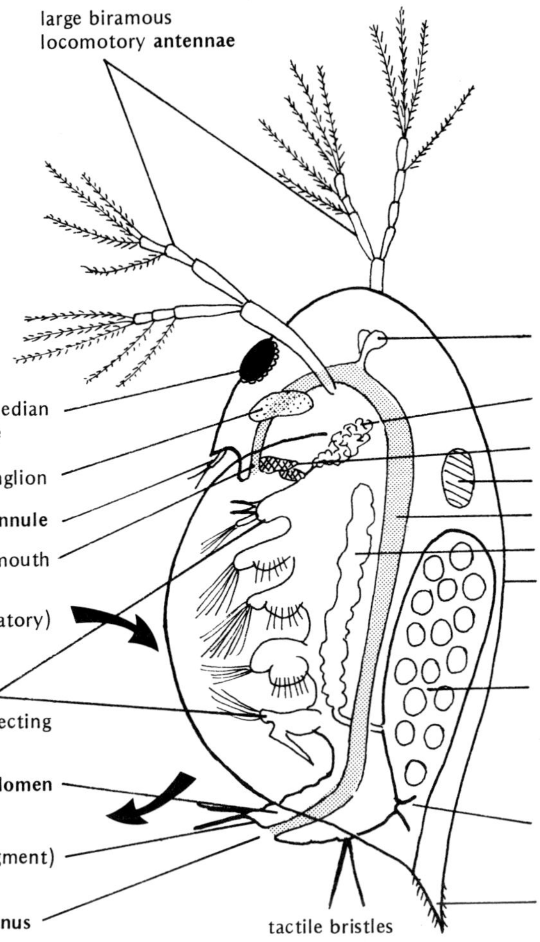

External features, parthenogenetic female
(x40. May reach 2mm in length)

Notes

1. Body laterally compressed.
2. Blood plasma contains haemoglobin.
3. Locomotion by movements of large antennae, resulting in jerky swimming.
4. Inhabits freshwater ponds and ditches. Very common.
5. Filter feeding – stream of water drawn into carapace space by movements of thoracic appendages – food particles in it are trapped by the fine bristles fringing the appendages – food particles then moved forwards towards mouth along mid-ventral food groove and eventually pushed into mouth by mandibles.

Reproduction

In favourable conditions female offspring are produced prolifically (up to 100 every 3 days) from unfertilized eggs, e.g. parthenogenetically.

This continues for successive generations until low temperatures, lack of food, etc. cause some eggs to develop into males. These conditions also induce the females to produce fewer, larger eggs which must be fertilized to develop further. The lining of part of the brood pouch thickens to form a box to house these eggs. It becomes detached at next moult and protects the eggs from freezing and desiccation until they hatch into parthenogenetic females when conditions become favourable again.

Prolific reproduction whenever possible is essential to compensate for heavy losses due to predation. Water-fleas provide the major food source for Hydra and for many fishes, etc.

Phylum Arthropoda
Class Crustacea
Sub-class Malacostraca
Order Isopoda

SUB-CLASS MALACOSTRACA

Order Isopoda

General characteristics

Body flattened *dorso-ventrally*.

No carapace.

Compound eyes are sessile.

Thorax composed of 8 segments, bearing 7 pairs of *similar appendages* (hence 'isopod').

Abdomen composed of 7 segments, most bearing respiratory appendages.

Development direct – eggs and developing embryos retained in a *brood pouch.*

e.g. *Oniscus asellus* (wood-louse)

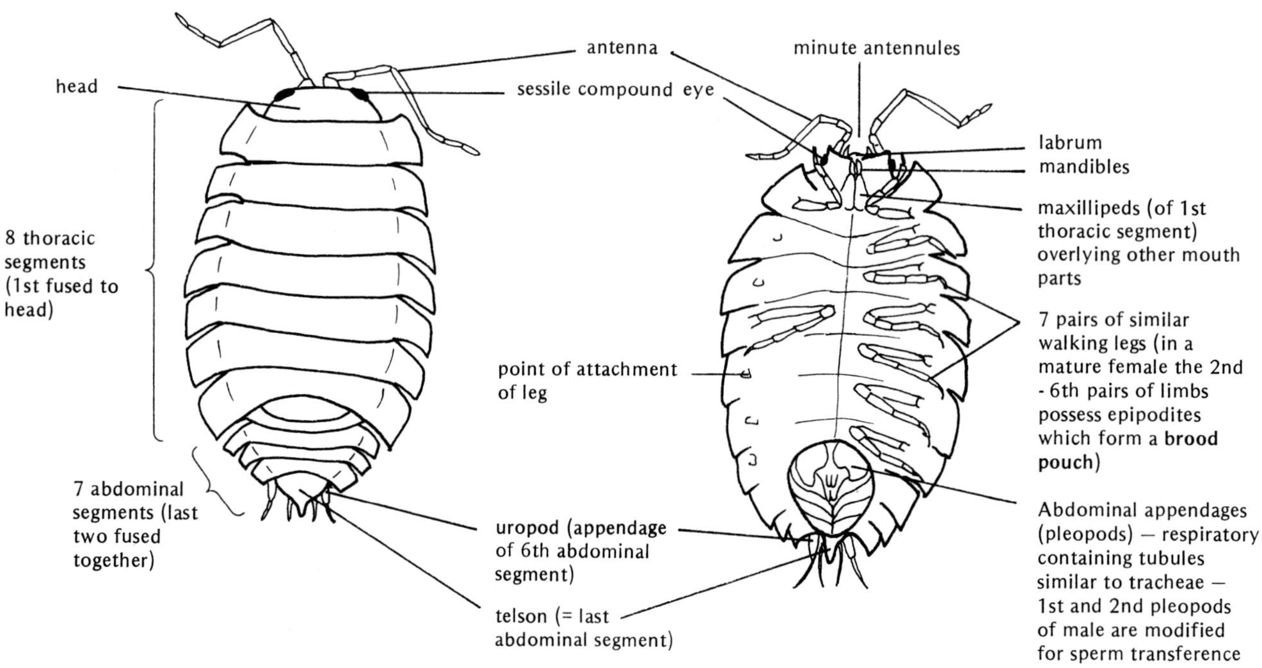

Dorsal view (x5)

Ventral view – male (5 legs removed)

Notes

1. The wood-louse is one of the few terrestrial crustaceans.
2. It is usually found in damp habitats.
3. It feeds on young leaves and rotting material.

Order Decapoda

Phylum Arthropoda
Class Crustacea
Sub-class Malacostraca
Order Decapoda

General characteristics

8 pairs of thoracic appendages: 3 pairs of maxillipeds, providing additional mouth parts to facilitate feeding – 5 pairs of walking legs (hence 'decapod').

Carapace present.

Compound eyes stalked and movable.

Abdomen composed of 6 segments and a telson.

5 pairs of pleopods typically present, usually modified to form swimmerets.

Sixth abdominal segment bears a pair of uropods, which together with the telson form the tail fan.

e.g. *Astacus pallipes* (crayfish)

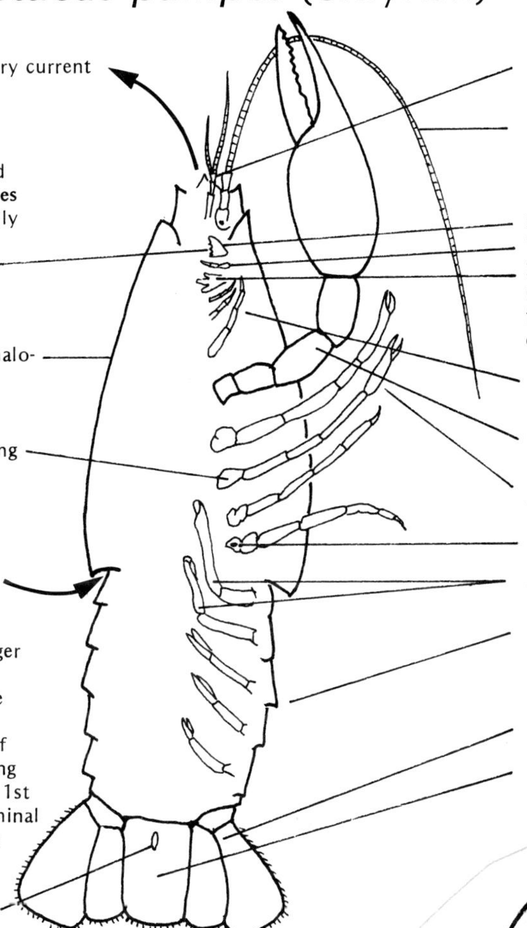

Ventral view – male (x1)
(Right-hand appendages removed)

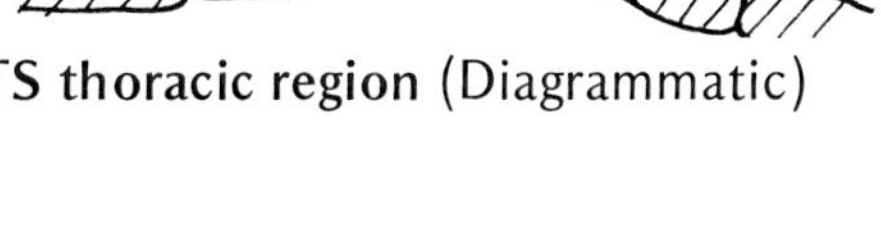

TS thoracic region (Diagrammatic)

Notes

1. The crayfish is found in freshwater streams and rivers which contain a high concentration of calcium salts.
2. It hides in crevices or burrows during the day, and feeds at night on any organic matter, living or dead.
3. The abdomen can be flexed rapidly, causing the animal to shoot backwards in the water if alarmed.
4. The sexes are separate.
5. Mating occurs in the autumn; eggs are fertilized externally and become attached to the pleopods of the female. They hatch in spring as young crayfish, which cling on to the pleopods for some time until all their appendages are fully developed. (Direct development is unusual in Crustacea.)

CLASSES DIPLOPODA & CHILOPODA

(formerly both included in Class Myriapoda)

Class Diplopoda (millipedes)	**Class Chilopoda (centipedes)**
Terrestrial – dark, damp habitats	Terrestrial – dark very damp habitats
Cylindrical segmented body	Flattened segmented body
Head bears: 1 pair antennae 2 pairs mouth parts simple eyes (ocelli)	Head bears: 1 pair antennae 3 pairs mouth parts simple eyes
Thorax of 4 segments, 3 of which bear 1 pair of walking legs. Abdomen of indefinite number of segments fused in pairs, so that each apparent 'segment' bears 2 pairs of walking legs	Body of indefinite number of segments – no division into thorax and abdomen. First body segment bears 1 pair poison claws, remaining segments each bear 1 pair walking legs
Up to 200 pairs of short legs	15 to 20 pairs of long slender legs
Hard cuticle, strengthened by lime salts	Elastic flexible cuticle
Tracheal system simple and unbranched – spiracles obscure	Tracheal system branched – spiracles permanently open (humidity essential for survival)
Excretion by Malpighian tubules opening into hind gut	Excretion by Malpighian tubules opening into hind gut
Vegetarian scavengers, feeding mainly on decaying plant material, but also on roots and fallen fruit	Active carnivorous predators, feeding mainly on insects and worms
Sluggish; glide along. May burrow	Fast moving; wriggle along
Coil up for protection; also emit noxious smell from stink glands along side of body	Poisonous claws for defence and capture of prey
Sexes separate. Genital pores on 3rd thoracic segment.	Sexes separate. Genital pores at posterior end of body.
Reproduction in Spring/Summer. Numerous eggs deposited in cavity dug by female. The nest is guarded.	Reproduction in Spring/Summer. Sometimes numerous eggs are laid in a cavity and guarded. Or eggs are laid singly and abandoned (as in Lithobius)
Development direct, but young have fewer segments.	Development direct, but young may have fewer segments.
E.g. Julus (millipede)	E.g. Lithobius (centipede)

Phylum Arthropoda
Classes Chilopoda
and Diplopoda
(formerly Myriapoda)

Millipede

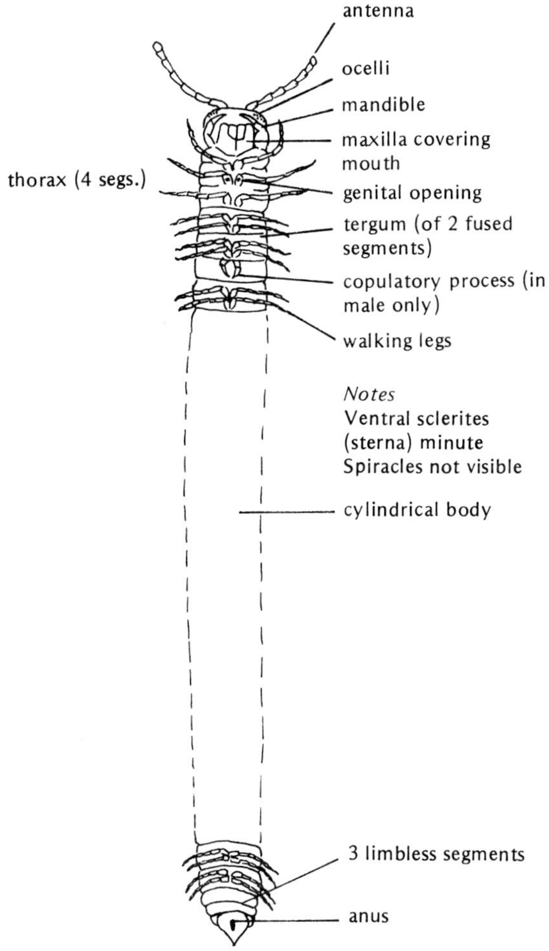

Julus terrestris – ventral view (x5)

Centipede

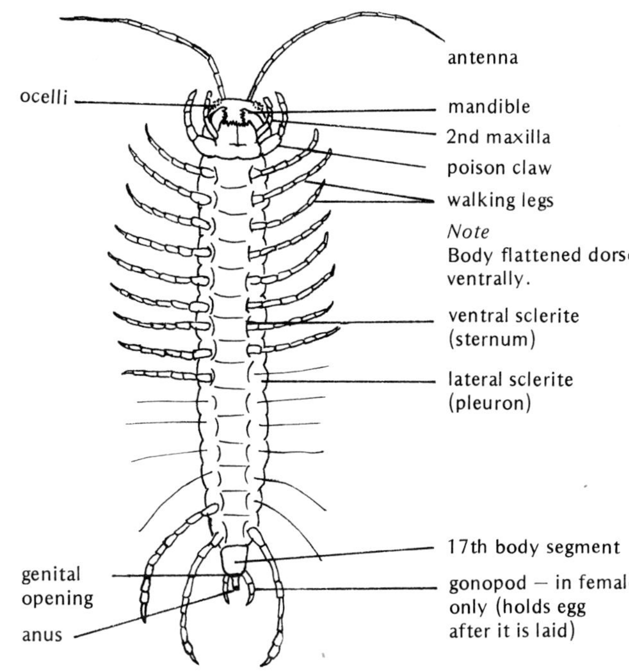

Lithobius – ventral view (x3)

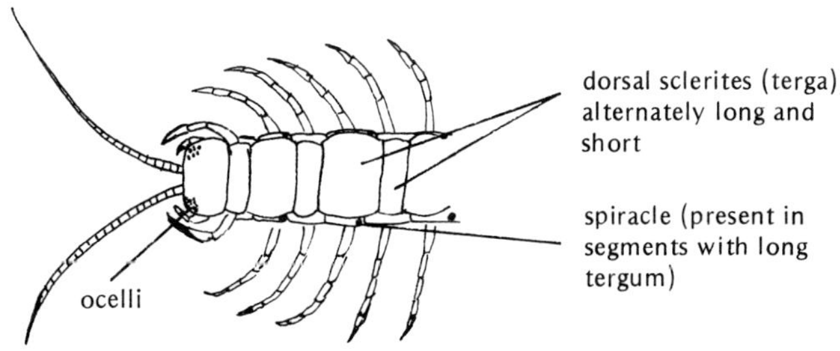

Lithobius forficatus – dorsal view (x3)

Phylum Arthropoda
Class Arachnida

CLASS ARACHNIDA

General characteristics

Typically terrestrial, predaceous carnivores. Commonly ectoparasitic (e.g. ticks and some mites).

No antennae, no mandibles, no compound eyes.

Abundant sensory bristles.

Body divided into cephalothorax (*prosoma*) and abdomen (*opisthoma*).

Cephalothorax bears: 1 pair of small *chelicerae*, often pincer-like – 1 pair of *pedipalps* which are often very large and may be grasping, sensory or locomotory – 4 pairs walking legs – simple eyes.

Abdomen bears no appendages except spinnerets in spiders.

Only liquid can be ingested (except in some mites) – food partially digested externally.

Typical respiratory organs are lung books, but a poorly-developed tracheal system may also be present.

Sexes separate. Development direct.

Examples of this Class are spiders, scorpions, ticks, mites and harvest-men.

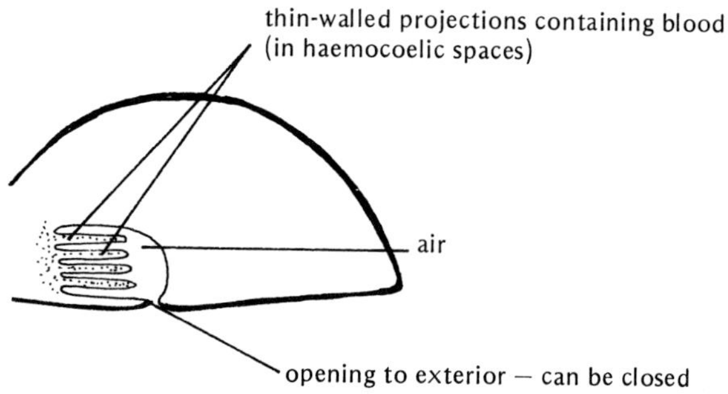

Diagram of lung book

e.g. *Araneus diadematus* (garden-spider)

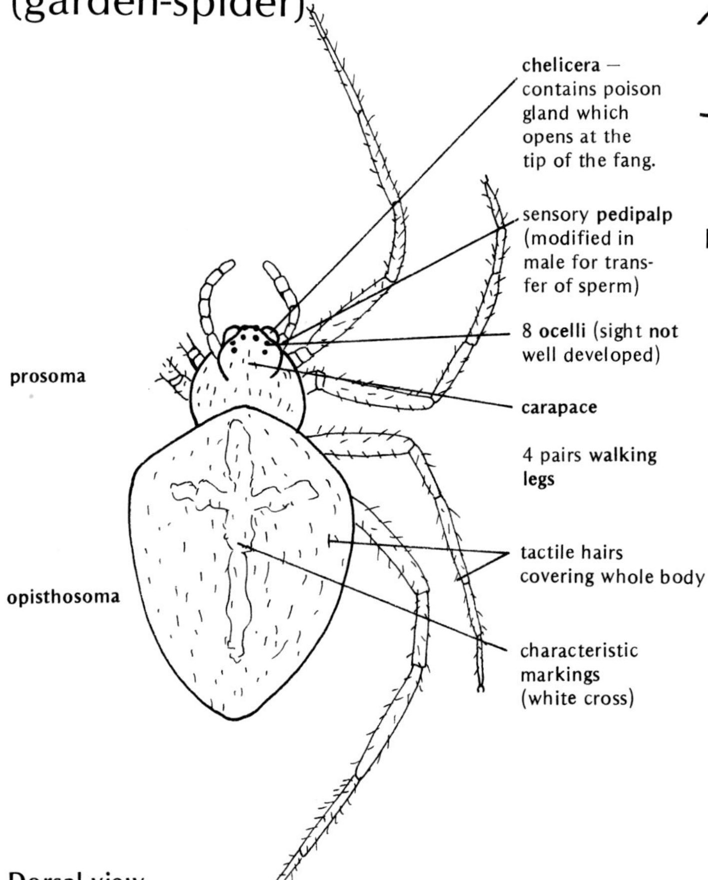

Dorsal view
Female garden spider (x5)

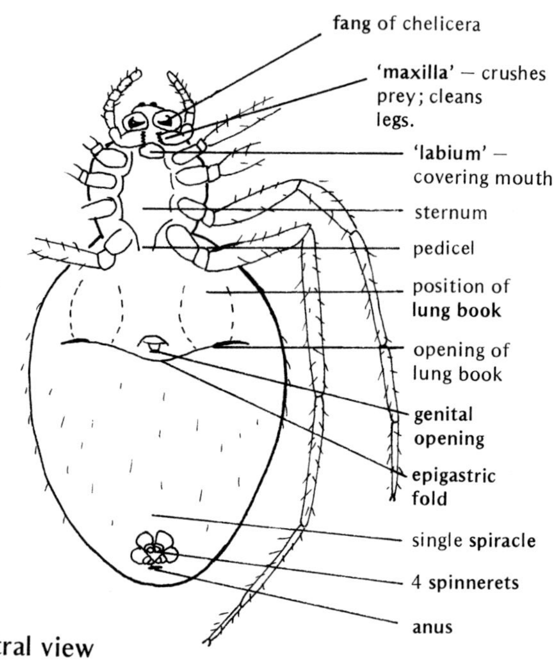

Ventral view

Notes

1. The spider feeds mainly on insects trapped in its web. The prey is paralysed by poison and then liquefied by secretions before being sucked up by means of a sucking stomach. Food can be stored in pouches off the digestive tract, so that the spider can go for long periods without food. However, it needs water to drink.
2. The male spider is much smaller than the female. When sexually mature he spins a small web on which he deposits a drop of seminal fluid. After dipping his pedipalps in this, he goes in search of a female. Courtship takes place on her web, and copulation involves the insertion of the male pedipalps into the genital opening. The male is often killed and eaten by the female after mating. Sperm can be stored by the female, and used to fertilize several batches of eggs. The eggs, laid in September or October, are protected within a silken cocoon. Development is direct, the young spiders hatching in May.

CLASS INSECTA

This class is larger in numbers of species and of individuals than any other group of terrestrial animals.

General characteristics

Primarily terrestrial, and typically adapted for flying but some species live in fresh water for part or whole of life history. No marine species.

Respiratory system of internal air tubes, i.e. tracheal system (even in aquatic forms). Paired segmental spiracles typically present.

Exoskeleton strong but light; chitin hardened by proteins.

Body usually clearly divided into head, thorax and abdomen. Sensory hairs, bristles, papillae.

Head derived from 6 segments. Bears 1 pair antennae, 3 pairs mouth parts. (Second pair of maxillae fused to form labium) and 1 pair compound eyes. Colour vision.

Thorax of 3 segments (pro-, meso-, and meta-) each bearing 1 pair of *walking legs.* Meso- and metathoracic segments each typically bear 1 pair of wings.

Abdomen primitively of 11 segments – fewer in more advanced insects. Bears no locomotory appendages. Gential appendages on eighth and ninth segments, sensory anal cerci commonly present on last segement.

Excretion by Malpighian tubules, opening into hind gut.

Blood colourless.

Sexes separate.

Development usually involving metamorphosis (either complete or incomplete). Growth in size occurs in stages (instars), the cuticle being shed between successive instars.

All insects are restricted to a small size, due to the problems imposed by the weight of the exoskeleton, and also due to the nature of their respiratory system.

Sub-class Apterygota	**Sub-class Exopterygota**	**Sub-class Endopterygota**
Primitively wingless	Wings develop externally	Wings develop internally
Body not clearly divided		
No metamorphosis – i.e. young similar to adults except in size and maturity	Incomplete metamorphosis – young = nymph, resembling adult but lacking wings. Develops gradually by series of moults, wings developing by small increments	Complete metamorphosis – series of similar larval instars unlike adult, then abrupt change to pupal instar and finally fully grown sexually mature imago
Young and adults share same habitat and exploit the same food	Nymphs and adults often share same habitat and exploit the same food	Larva and adult exploit different habitats
E.g. the following genera: **Tomocerus** (springtail) **Lepisma** (silverfish)	**Periplaneta** (cockroach) **Locusta** (locust) **Carausius** (stick insect) **Libellula** (dragonfly) **Termes** (termite) **Forficula** (earwig) **Aphis** (greenfly)	**Pieris** (white butterfly **Apis** (bee) **Musca** (house fly)

The tremendous success of insects is partly due to the great amount of external variation shown by the group. The insect body plan lends itself readily to structural specialisation for almost any mode of life, and thus permits an extraordinary degree of adaptive radiation.

Phylum Arthropoda

Class Insecta

The advantages of a larval stage in the life history of insects

The nymphs of Apterygote and Exopterygote insects are usually similar to the adults in mode of life and in general appearance. Since they possess a rigid exoskeleton of hardened chitin, increase in size during growth involves the shedding of the exoskeleton at frequent intervals; it can be stretched only to a very limited extent, at the joints.

Ecdysis is not only a difficult operation, but it exposes a new *soft* cuticle, and until this hardens the nymph is very vulnerable; it cannot move efficiently, and is pale in colour and less well camouflaged. It must therefore hide until its skeleton hardens.

The inclusion of a larval stage in the life history, as in the Endopterygota, overcomes this problem. The larva is soft bodied and of simple form; its cuticle can be stretched appreciably and is relatively easy to shed. But a soft skeleton prohibits rapid movement, and so the mode of life of a larva tends to be quiet and unobtrusive. It is usually well camouflaged and spends its time merely feeding on food which is abundant and close at hand, and which is usually different from the food of the adult. Metamorphosis occurs suddenly during a pupal stage, during which the insect is protected from its surroundings by a rigid pupal skin. A fully grown, sexually mature *imago* emerges from the pupa.

Thus the larval stage endows the species with many advantages and confers much more flexibility to the Endopterygota as a whole:

i. The process of growth is less hazardous.

ii. Competition between the young and the adults is eliminated; their modes of life are quite different.

iii. Different food sources are used by the larva and the adult, and thus seasonal fluctuations can be exploited, e.g. leaves are eaten by caterpillars in the spring, and nectar by butterflies in the summer.

iv. The pupal stage ensures survival in severe conditions.

v. Since the adult does not moult its body form is open to greater complexity, particularly the wings.

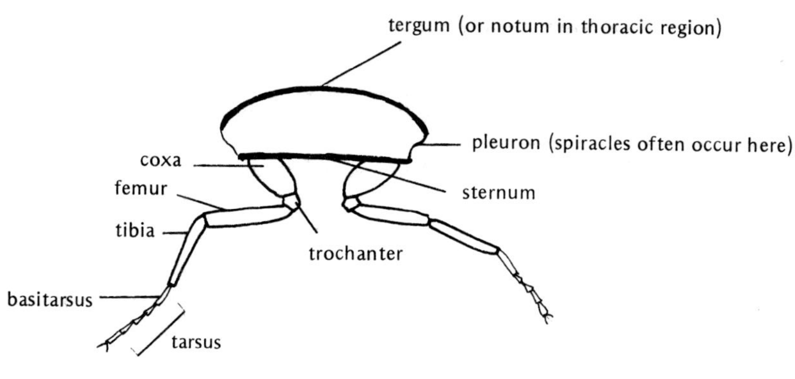

TS thorax of generalised insect to indicate terminology

SUB-CLASS APTERYGOTA

Phylum Arthropoda
Class Insecta
Sub-class Apterygota

General characteristics

Wingless

Body not clearly divided into head, thorax and abdomen.

No metamorphosis.

e.g. *Lepisma saccharina (domestic silverfish)*

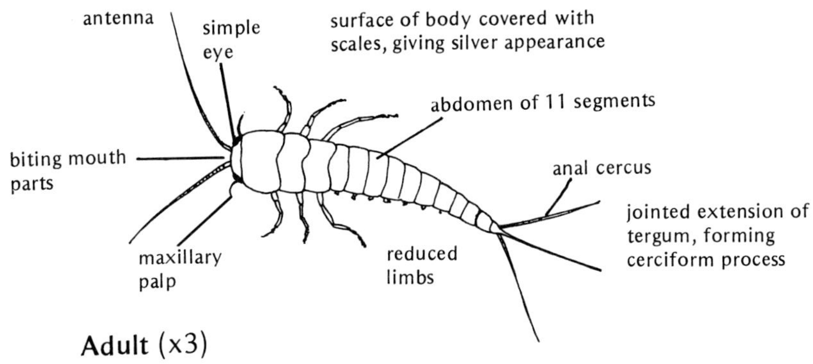

Adult (x3)

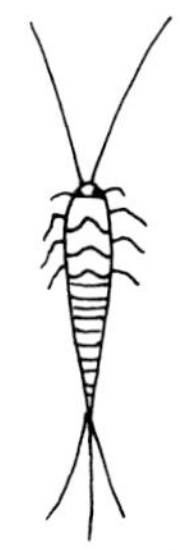

Nymph

SUB-CLASS EXOPTERYGOTA

Order Dictyoptera (most primitive group of winged insects)

Phylum Arthropoda
Class Insecta
Sub-class Exopterygota
Order Dictyoptera

General characteristics

Biting mouth parts.

All legs similar, modified for running.

Forewings thickened to form elytra (or tegmina) covering membranous hindwings.

e.g. *Periplaneta americana* (cockroach)

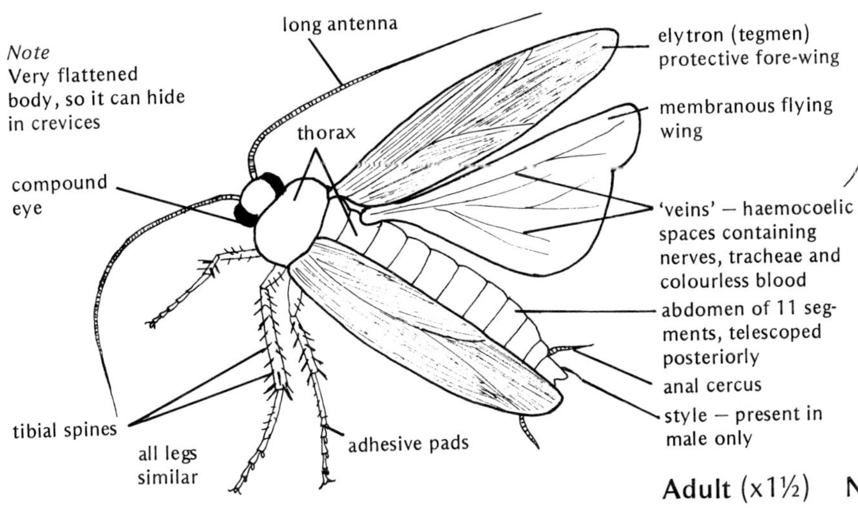

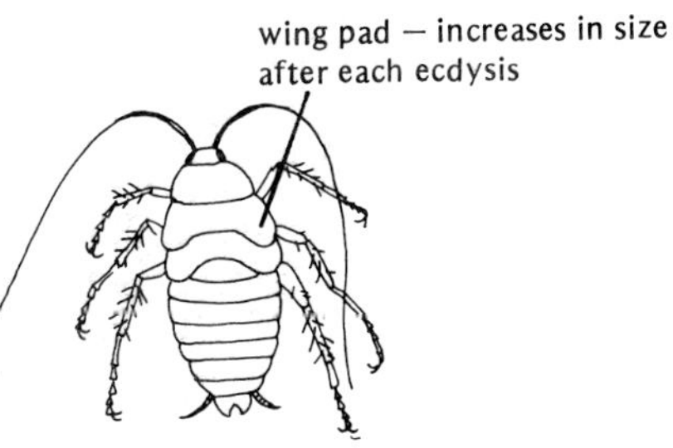

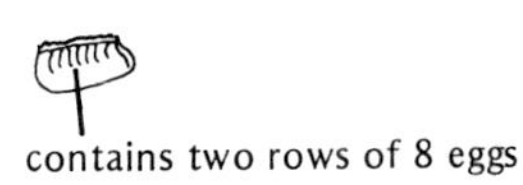

Adult (x1½) Nymph (x2) Egg case (x1)

Notes

1. Cannot live out of doors in Britain, but can set up infestations in kitchens, etc. of warm buildings.
2. Hide by day, scavenge by night.
3. Run rapidly, rarely fly.
4. Breeding can occur throughout the year. Eggs enclosed in purse-like ootheca which is carried protruding from the end of the abdomen of the female for about a week (while the case is still soft) before being deposited. Young emerge as nymphs which take 1 year to develop, moulting 7 times.

Phylum Arthropoda
Class Insecta
Sub-class Exopterygota
Order Orthoptera

Order Orthoptera (locusts, grasshoppers, crickets)

General characteristics

Generally large.

Usually herbivorous.

Hind legs modified for jumping.

Stridulatory organs often present in male.

Auditory organs present.

e.g. *Locusta migratoria* (migratory locust)

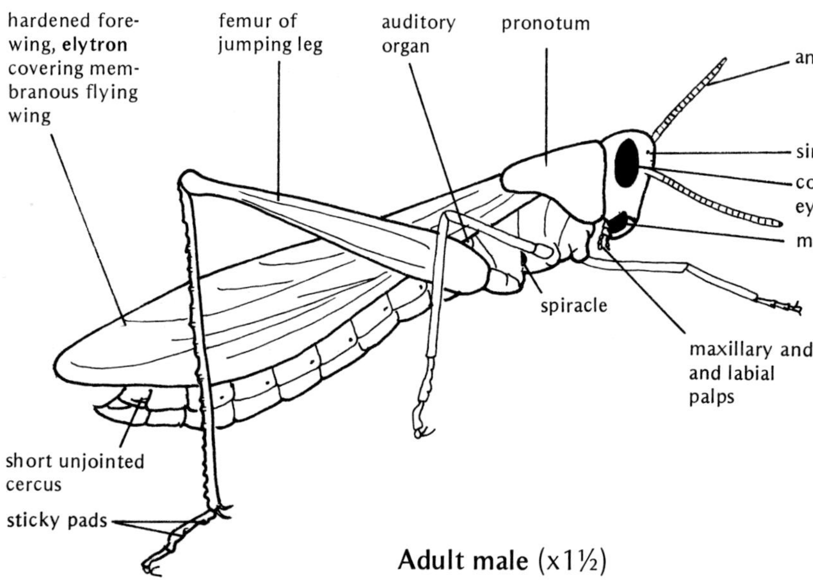

Adult male (x1½)

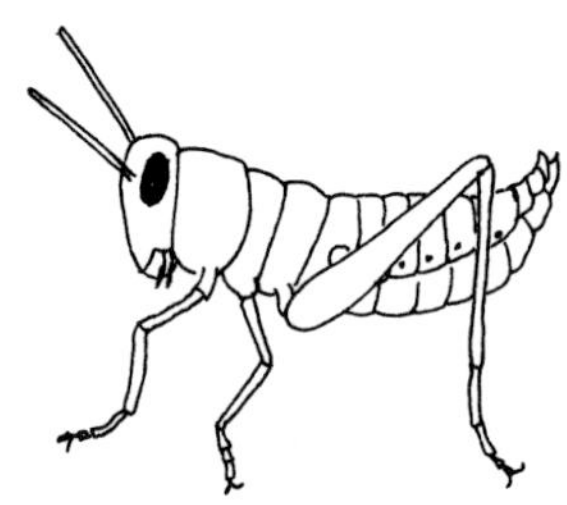

Nymph – Hopper (x3)
(Early instar – wing pads start to develop after the second moult)

Notes

1. Occur only in hot countries. Warm, sandy substratum needed in which to lay eggs. Female inserts abdomen deeply into soil and lays 'pods' of eggs in batches of about 100. Worm-like pronymphs hatch from the eggs and wriggle to the surface where they moult. The nymphs moult 5 more times to become adult in about 6 weeks.
2. Plagues of locusts arise after a period of good conditions has resulted in a large population increase and over-crowding. Young which are subjected to crowded conditions during growth develop into gregarious individuals which differ in appearance and behaviour and migrate in swarms to new areas.
3. Male has a roughened ridge on the inner side of each femur of the hind leg. This can be rubbed against a hard prominent 'vein' on the elytron to produce a strident noise.

Order Odonta (dragonflies)

Phylum Arthropoda
Class Insecta
Sub-class Exopterygota
Order Odonata

General characteristics

Strong fliers – 2 pairs of large membranous wings, equal in size.

Predaceous carnivores – powerful biting mouth parts.

Enormous compound eyes.

Aquatic predaceous nymph with specialised prehensile labium.

e.g. *Libellula sp.* (broad-bodied dragonfly)

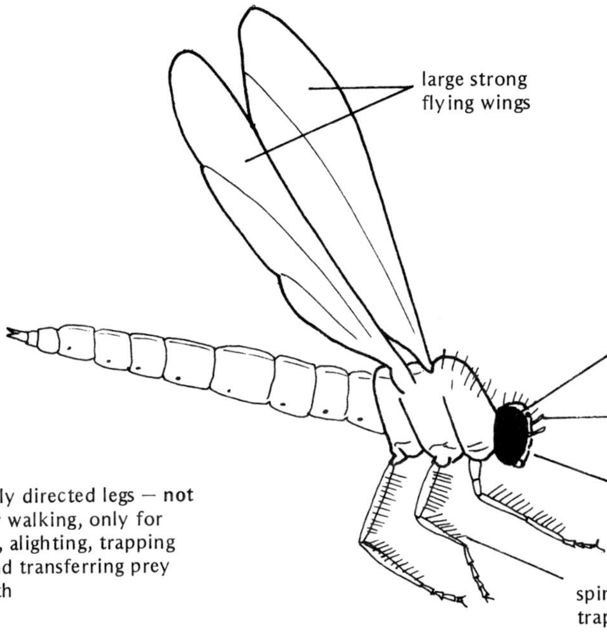

Adult – female (x1½)

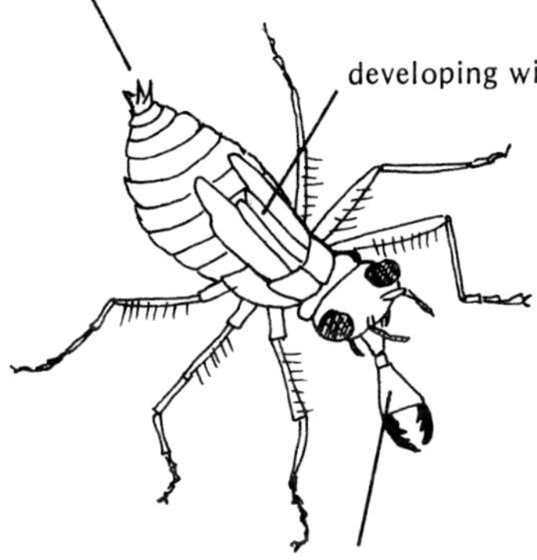

Aquatic nymph – last instar (x1)

Notes

1. Mating takes place on the wing.
2. The eggs are laid in winter, the female dipping the end of her abdomen into the water during flight to wash the eggs off the tip. They sink to the bottom.
3. Nymph lives 2 years or more amongst stones and weeds at the bottom of a pond, half-buried in mud. It preys on other larvae and, later, on tadpoles and even small fish. Climbs out of water for final moult.
4. Adult is diurnal and sun-loving, and lives for only a few months.

Phylum Arthropoda
Class Insecta
Sub-class Endopterygota
Order Lepidoptera

SUB-CLASS ENDOPTERYGOTA

Order Lepidoptera (moths and butterflies)

General characteristics

2 pairs of large membranous wings, colourfully patterned by cuticular scales (modified hairs).

Fore and hind wings can be interlocked by hooks.

Adults are liquid feeders, food being sucked up by a proboscis (modified maxillae).

No mandibles.

Larva polypodous (many legs); known as a caterpillar. Usually herbivorous.

e.g. *Pieris brassica* (large cabbage white butterfly)

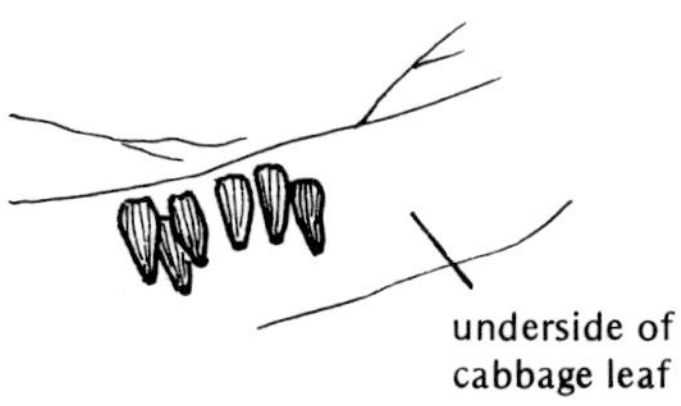

Eggs (x4)

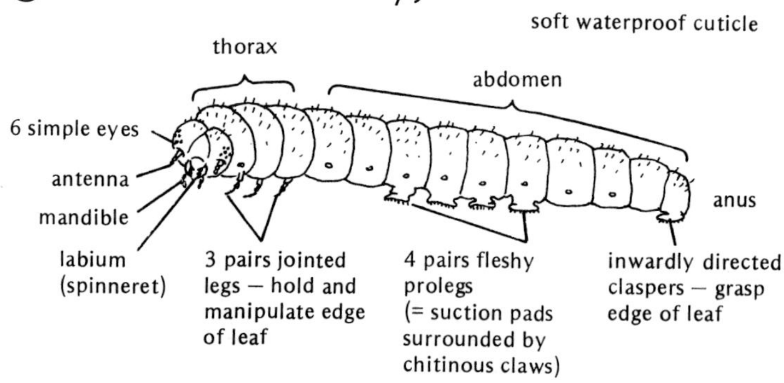

Full grown larva – 8 weeks old (x2½)

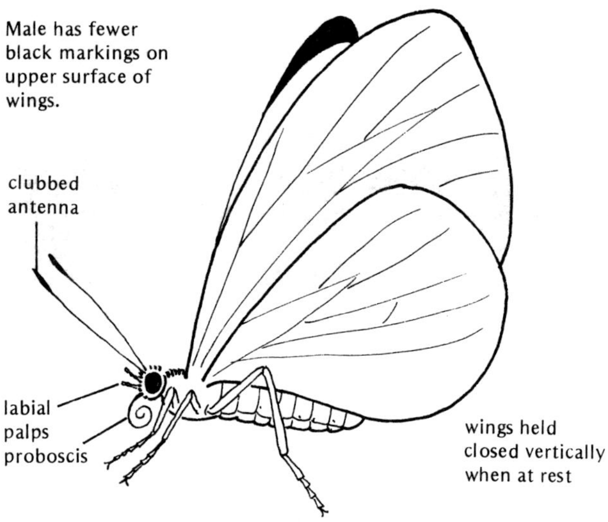

Imago – fully grown, sexually mature (x1½)

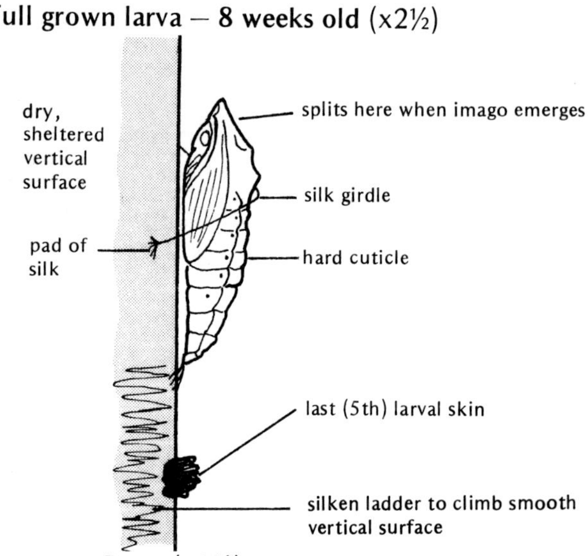

Pupa (x1½)

Notes

1. Eggs are laid in clusters on the underside of a cabbage leaf in May and August. They hatch after 1 week. The larvae attach themselves to the leaf with silken thread, feed voraciously and grow rapidly. They move to the edge of the leaf after the third moult.
2. Pupation takes 2 to 3 weeks but the imago of the second brood does not emerge until the winter is over.
3. Imago feeds only on nectar. Mating takes place on the wing, male courting female by shedding scented scales over her.
4. Unpalatable to birds, therefore cryptic colouration unnecessary (i.e. no need to be camouflaged).

Order Diptera (flies, mosquitoes)

Phylum Arthropoda
Class Insecta
Sub-class Endopterygota
Order Diptera

General characteristics

Only 1 pair of membranous wings, the hind pair of wings having been reduced to halteres which control the equilibrium of the flying insect.

Suctorial mouth parts, no mandibles.

Legless (apodous) larva.

Pupa usually enclosed by last larval skin (puparium). No cocoon.

e.g. *Musca domestica* (house-fly)

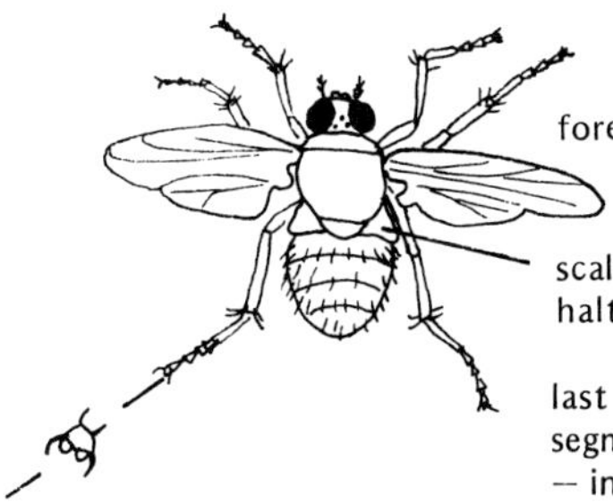

Imago (x3)

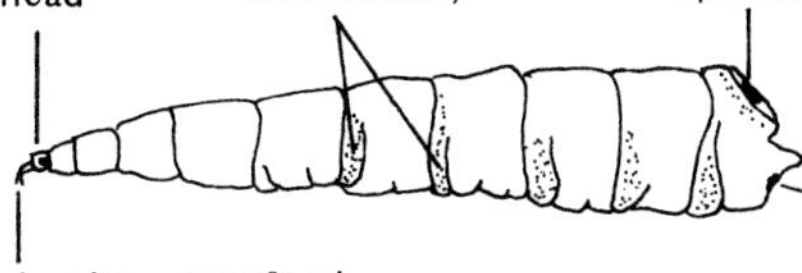

Full grown larva (Maggot) (x4)
(white in colour)

Side view of head of feeding fly (Very enlarged)

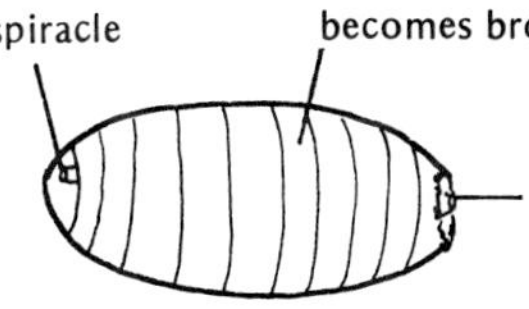

Puparium containing pupa (x4)

Notes

1. Only liquid food can be ingested. Saliva passes via the food canals of the labellum on to the food which becomes semi-digested and can then be drawn along the food canals and up the oesophagus (by muscles in the proboscis). Partly digested food is often regurgitated later as more saliva is added to it. This results in 'vomit spots'.
2. House-flies feed on almost any organic matter, including faeces, human foods and decaying detritus. Their indiscriminate feeding habits are responsible for the spread of many diseases including typhoid, cholera, diarrhoea and dysentery.
3. Eggs are laid in summer in batches of 100 to 150 within decomposing organic matter. The larvae hatch after 8 to 24 hours, depending on temperature. Moulting occurs twice, and the maggot is fully grown and about 1 cm long after about 5 days.
4. Until fully grown the maggot moves away from light, remaining in the warm, dark, moist regions of its habitat. Finally it migrates to a dryer, cooler region and pupates. The imago emerges after about 3 days to one week, the puparium having been burst open by means of a ptilinum. This is a flexible part of the head capsule which, until the cuticle hardens, can be distended with blood.
5. The whole life cycle may be completed in 3 weeks, and there may be 9 generations in a year.
6. Most house flies die in late autumn, though a few may hibernate indoors. Survival during the winter is usually in the pupal stage.

Phylum Arthropoda
Class Insecta
Sub-class Endopterygota
Order Hymenoptera

Order Hymenoptera

General characteristics

2 pairs of membranous wings, interlocked by hooklets during flight.

Mouth parts enable biting, licking and sucking.

Ovipositor often modified for piercing/sawing/stinging.

Larva apodous. Pupa enclosed in a cocoon.

Includes many genera highly specialised for social life.

e.g. *Apis mellifica* (honey-bee)

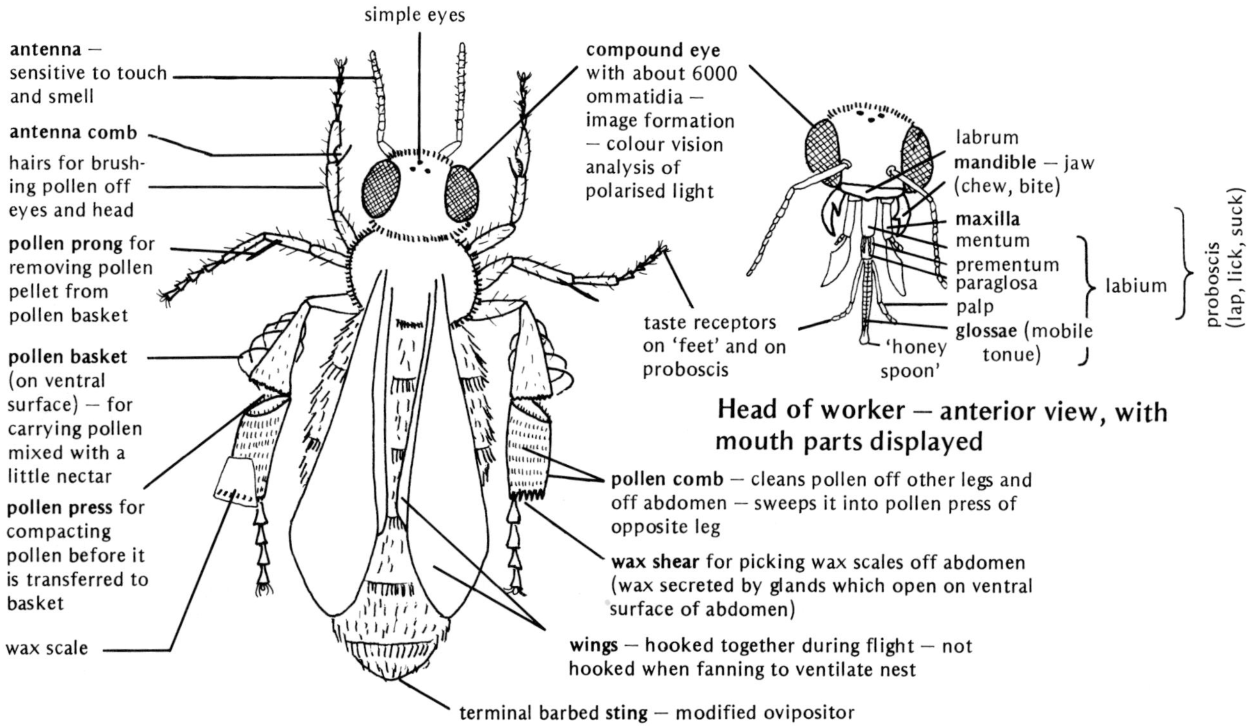

Head of worker — anterior view, with mouth parts displayed

Dorsal view of worker honey-bee — sterile female (x10)
(Note the highly specialised mouth parts and legs, enabling the bee to perform many varied tasks.)

Queen

Develops from fertilized egg, laid in a specially large queen cell.

Fed solely on royal jelly for entire life.

Twice the length of a worker.

Drone

Develops parthenogenetically, i.e. from unfertilized egg.

Enormous compound eyes.

Long, strong wings (longer than abdomen).

Stout and ungainly.

Feeble mouth parts; cannot feed himself.

No pollen baskets, etc.

Very sensitive antennae.

No sting.

Lazy and languid except when virgin queen emerges for her nuptial flight.

Phylum Arthropoda
Class Insecta
Sub-class Endopterygota
Order Hymenoptera
Genus *Apis*

Development of worker bee

Fertilized egg
Hatches after 72 hrs., larva fed only on royal jelly for 3 days

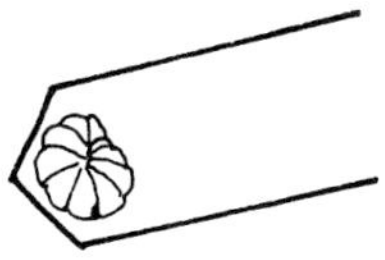

Larva
Third day after hatching – fills base of cell – is now fed on bee bread and honey

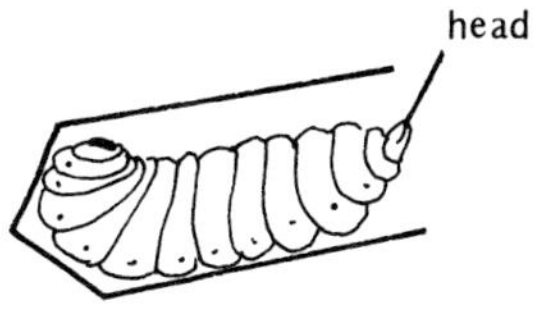

Larva – full grown
Fifth day – fills cell – spins cocoon

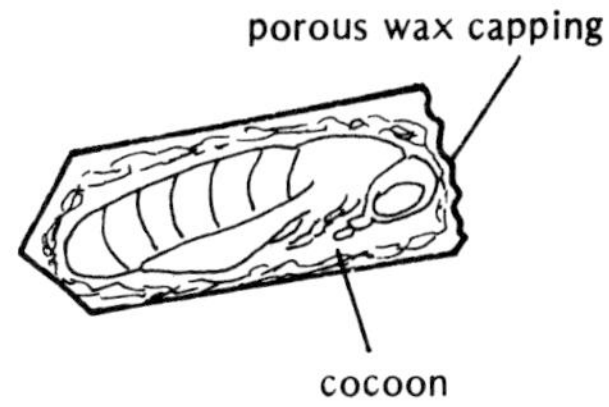

Pupa
Pupates for 8 to 10 days gradually becomes pigmented; imago emerges 21 days after larva hatches from egg

Notes

1. Honey-bees are social insects which form colonies consisting of a single mated egg-laying female (Queen), a 'worker' caste of up to 60,000 subordinate infertile females (workers), and a few hundred males (drones).
2. Bees are entirely dependent on nectar and pollen for food.
3. The queen's main function is to lay eggs. This she does continuously, night and day, except in winter. She also produces a secretion (queen substance) which maintains the cohesion of the colony; without it the bees cease to function as a unit.
4. The workers forage for food and maintain the nest, the work being divided amongst them according to their physiological age. Thus, on emerging from the pupa, the worker at first remains in the brood chamber, cleaning cells then feeding the older larvae from the surrounding food storage cells (containing processed pollen and nectar). She then feeds the youngest larvae, and perhaps the queen, on royal jelly which she starts secreting from her brood food glands (specialised salivary glands). After these atrophy, her wax glands become productive and the worker then leaves the brood chamber to become involved in building and repairing the wax cells. After this she participates in general duties, such as cleaning the nest, helping with comb construction and repair, relieving the foragers of their loads and carrying the pollen and nectar to storage cells, processing it and storing it. At this stage she also makes orientation flights from the nest to learn the geography of the area. Then after a period of guarding the nest she finally (aged about 3 weeks) becomes a forager, spending the daylight hours seeking pollen, nectar, water and, especially in Autumn, propolis from sticky buds. Having found a rich source of food the bee is able to communicate its whereabouts to the other foragers by means of bee dances performed in the combs. Worker bees normally live for only 6 to 8 weeks, except those reared late in the year which overwinter in a 'cluster' with the queen, the nest having been prepared for cold weather by stopping up any cracks with propolis, and the drones having been driven out of the nest in the Autumn.
5. The nest consists of a series of combs hung vertically from the ceiling and reaching almost to the floor. They are made from beeswax and are composed of thousands of hexagonal cells arranged on both sides of the comb (i.e. back to back).
6. Most of the cells are identical and house either the eggs, larvae and pupae destined to become workers, or food stores (honey and processed pollen, 'bee bread'). Some cells are slightly larger; these house eggs that will produce drones; a few are enormous and acorn-shaped and are used for rearing queens. Whether a fertilized egg develops into a queen or a worker is determined by the diet during the larval stage – a larva must be fed solely on royal jelly to produce a queen. Worker larvae are fed on royal jelly for only the first 3 days, after which they are fed on honey and bee bread (the former supplies carbohydrates, the latter protein).
7. Colony increase is achieved by swarming; new queens are reared, but before the first of them emerges the old queen leaves the nest together with about half the colony. A new nest is constructed in a suitable site which has been previously selected by 'scout' workers. Meanwhile when the first virgin queen emerges in the original nest, the remaining developing queens are usually killed, the new queen is mated during a nuptial flight, and turns to head the colony.

15. Phylum Mollusca

General characteristics

Triploblastic and coelomate, but unsegmented.

Coelom very reduced (restricted to the pericardium and the cavities of the gonads and excretory organs).

Soft fleshy body divided into *head* with tentacles and eyes, ventral muscular *foot*, and dorsal *visceral mass*, overarched by a sheet of tissue known as the *mantle* which secretes a calcareous *shell.*

Well developed open blood system – blue respiratory pigment (haemocyanin) may be present. Respiratory organs typically = paired ciliated gills (ctenidia).

Sexes usually separate. Trochophore larva when indirect development occurs.

Basic bilateral symmetry with two growth axes (since mantle grows by marginal increase), but asymmetry common in gastropods, due to torsion of the visceral mass.

Very successful phylum, with great number of species (second only to the arthropods). Exceptionally diverse in form and modes of life, but essentially marine, and conforming to a generalised body plan:

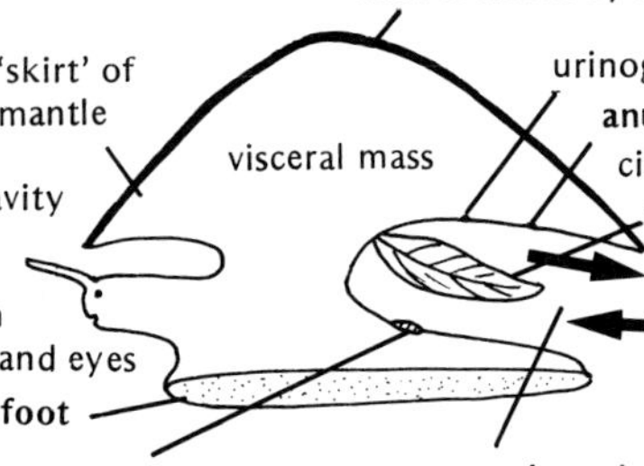

Class Gastropoda	**Class Bivalvia** (Lamellibranchiata)	**Class Cephalopoda**
Marine, terrestrial or freshwater	Mainly marine	Marine
Asymmetrical, due to 'torsion' of visceral mass, so that anus becomes anterior in position	Bilaterally symmetrical and bilaterally compressed	Bilaterally symmetrical
Well developed head with tentacles and eyes	Rudimentary head – no tentacles	Well developed head with cartilaginous cranium, brain, and highly developed eyes, comparable with vertebrates
Large flat foot for creeping	Small pointed foot for burrowing or boring into wood or rock for protection	Foot modified to form siphon, and long prehensile tentacles
Mantle secretes single shell, frequently coiled	2-lobed mantle secretes shell in two halves, completely enclosing body	Shell, if present, enclosed by mantle; mantle cavity modified for jet propulsion
Radula present	Filter feeding by enlarged ctenidia	Predacious carnivores with radula and horny jaws
Slow moving	Sedentary, with limited powers of locomotion	Active, fast moving. Very efficient circulatory system
Commonly hermaphrodite	Sexes separate, but may change during life	Sexes separate
Larva may or may not be present	Planktonic larva usually present	Direct development
E.g. the following genera:		
Helix (garden snail) **Patella** (limpet) **Littorina** (periwinkle)	**Mytilus** (mussel) **Cardium** (cockle) **Pecten** (scallop)	**Sepia** (cuttlefish) **Loligo** (squid) **Octopus**

CLASS GASTROPODA

Torsion and asymmetry

The visceral mass tends to be elongated in gastropods, and a coiled shell houses it in the most compact manner. *Torsion* of the visceral mass and mantle (i.e. rotation through 180°) brings the mantle cavity to the anterior. Here it can assume greater sensory importance, and also the pressure of weight of the viscera on the mantle cavity is relieved, the weight now resting on the posterior upper surface of the foot. But a major disadvantage of torsion is that the anus and excretory aperture open directly over the head. However, the displacement of these apertures to the right-hand side of the body is achieved by the asymmetric coiling of the shell, i.e. a pulling out of the spire of the shell, and a skewing upwards. This is accompanied by the loss of the organs of one side of the viscera and mantle cavity.

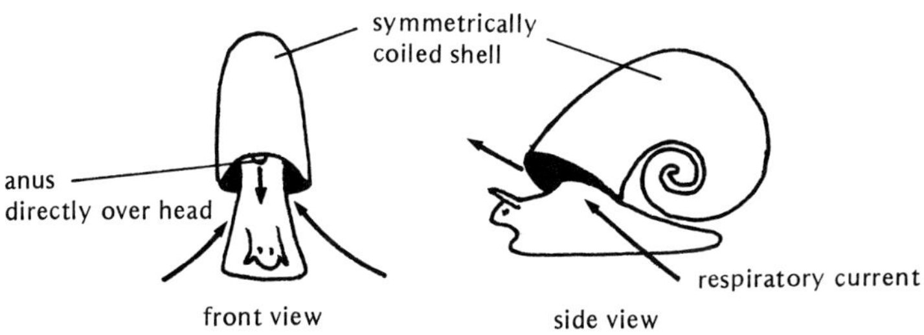

Theoretical primitive condition after torsion

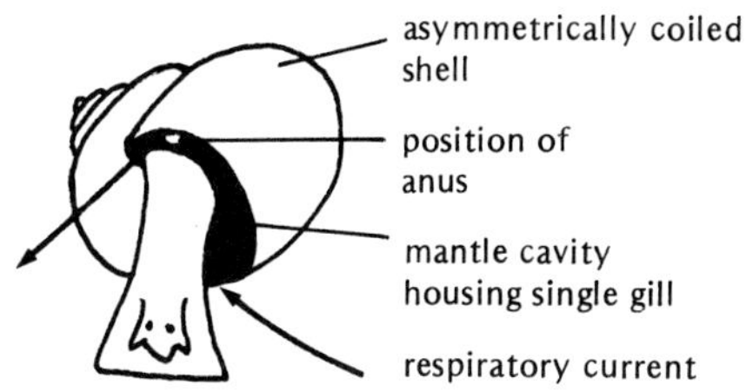

Torsion followed by asymmetric coiling of shell

e.g. *Helix aspersa* (garden snail)

External features (x2)

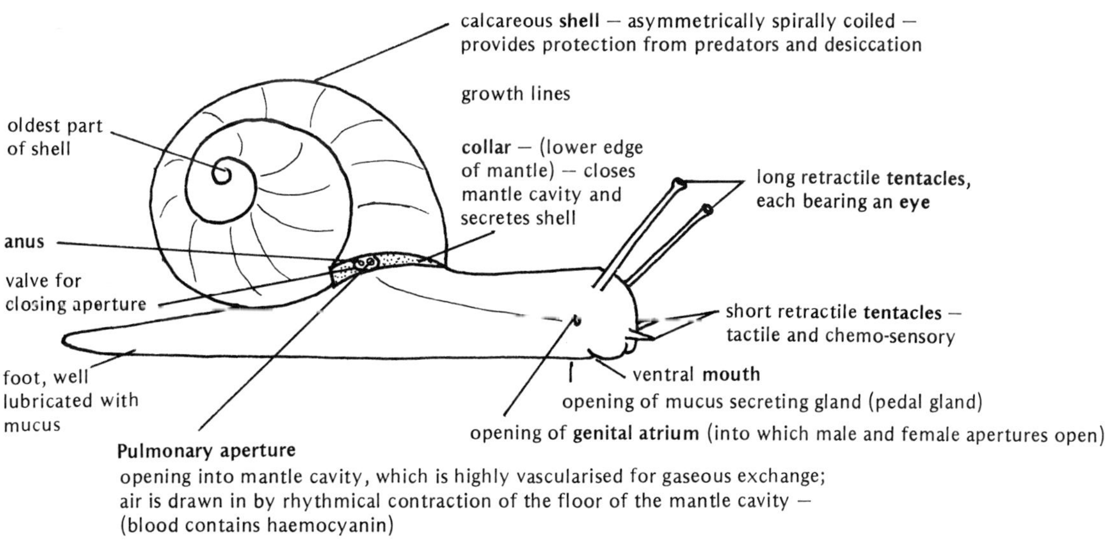

Notes

1. Terrestrial – herbivorous, feeding on living leaves and fruit, etc., which are fragmented by the toothed radula.
2. Active only in damp conditions, mainly at night – returns to fixed 'home' after foraging.
3. Hermaphrodite, but cross fertilization occurs involving complicated copulatory process – sperm can be stored – shelled eggs laid in summer – shells permeable (calcium phosphate), so that eggs must be laid in a damp situation – usually buried in the ground in batches.
4. Hibernates in winter – withdraws into shell, and closes aperture with mucous secretion hardened by calcium phosphate. Often burrows into ground to hibernate.

CLASS BIVALVIA

e.g. *Mytilus edulis* (common mussel)

External features (x1)

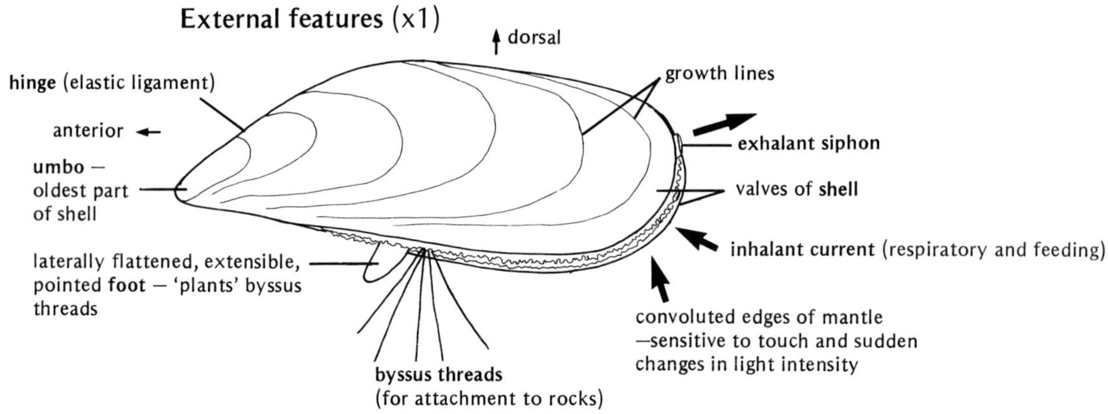

Notes

1. Marine — occurs, often in dense masses, between tide marks — tolerant of wide variations in salinity — attached to rocks, stones or wood by tough elastic byssus threads, secreted by glands at base of foot — the mussel can break the threads if necessary, either to make new longer ones as silt strained from the feeding current builds up beneath it, or to move, very slowly, to a new locality.
2. The ciliated gills are extensive. A powerful respiratory current is drawn through them by ciliary action, and at the same time food particles are strained from the water and directed to the mouth.
3. When the mussel is exposed at low tide, the 2 valves of the shell can be firmly closed, and anaerobic respiration then occurs.
4. The sexes are separate. Gametes are released into the sea usually early in the year; the eggs hatch into planktonic larvae.

CLASS CEPHALOPODA

e.g. *Loligo forbesi* (common squid)

External features (x $\frac{1}{3}$)

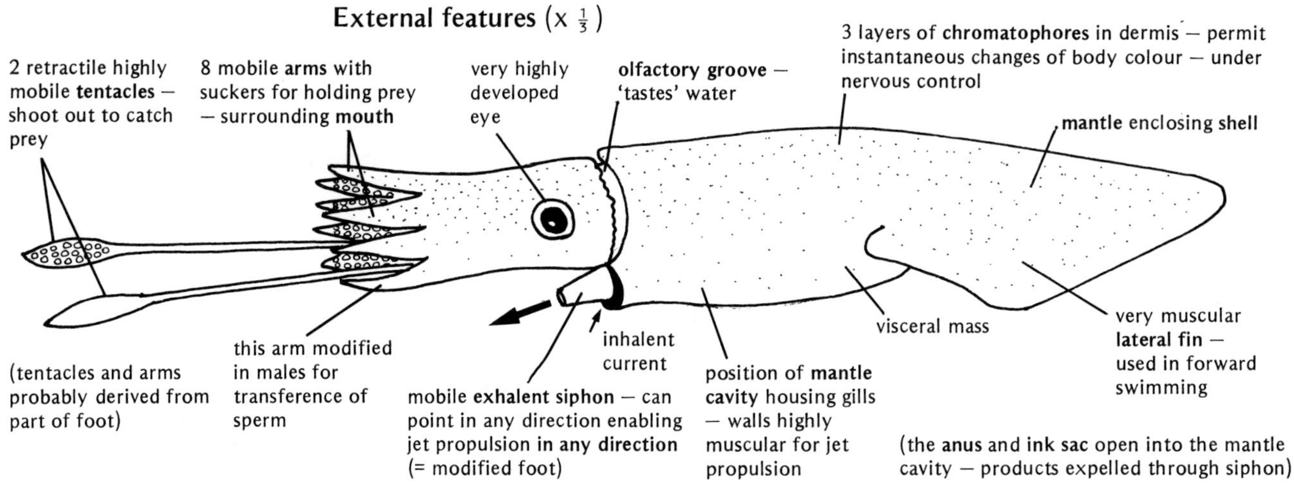

Notes

1. Marine — common in surface waters all around British coast, several miles offshore. Usually 20-30 cm long, but may reach 60 cm.
2. Feeds on crustaceans and small fish. The mouth is armed with horny jaws similar to a parrot's beak, and a radula is also present. Poison glands, for killing prey, open into the mouth.
3. Squids are the fastest of all aquatic animals, in short bursts of speed. During forward swimming the arms are closed together, so that the body is torpedo-shaped:

4. Protective colour changes can occur remarkably quickly (in less than 1 sec), and enemies (usually seals) can be evaded by emitting a cloud of black 'ink' (melanin).
5. The sexes are separate — fertilization occurs in the mantle cavity of the female where the eggs are enclosed in capsules and covered with a sticky secretion so that they can be attached to some support, usually in shallower waters. Development is direct.

16. Phylum Echinodermata

General characteristics

Unsegmented – basically bilaterally symmetrical, but radial symmetry (with 5 axes) acquired during development.

No head – central mouth, ventral in position.

No excretory organs. Rudimentary blood system.

Nervous system and sense organs not well developed, food detected mainly by chemoreception.

Mesodermal endoskeleton of calcareous plates, with movable spines protruding from surface of body (hence 'spiny-skinned').

Coelom extensive – divided into 2 distinct and separate parts; the perivisceral coelom associated with skin gills, and the unique water vascular system associated with locomotory tube feet.

Anus and opening of water vascular system (sieve plate or madreporite) aboral, opening on upper surface of body.

Exclusively marine. Slow moving.

Sexes separate, fertilization external; bilaterally symmetrical pelagic larva, the form and development of which indicates that the echinoderms are related to the same ancestral stock as chordates.

Section through arm of starfish to show general echinoderm structure (Gonads omitted)

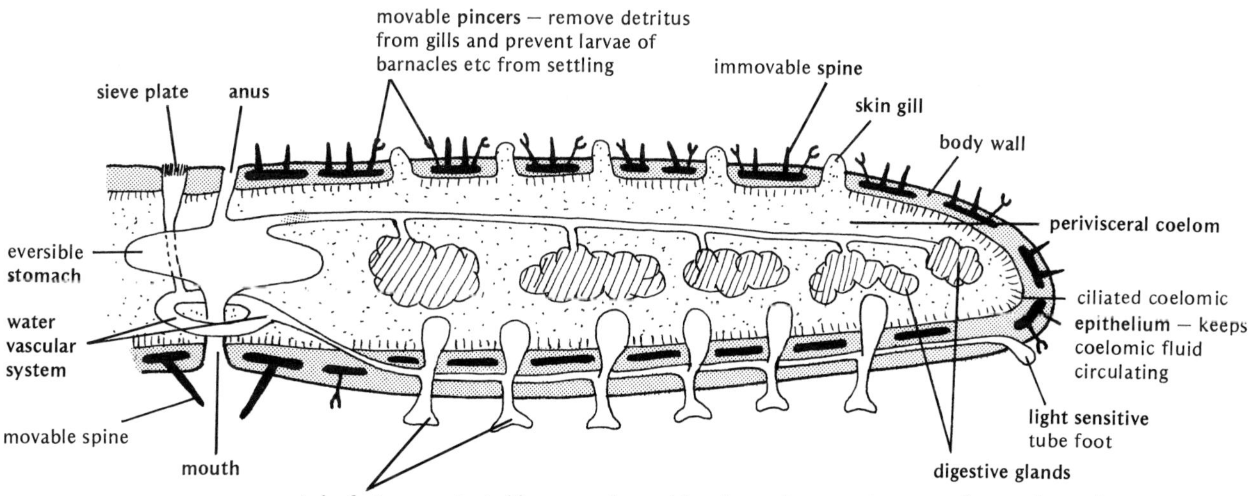

tube feet – protected by rows of movable spines – have great powers of expansion and contraction, fluid being pumped into them from the reservoirs. Each acts independently – terminates in a sucker – used for locomotion, firm attachment during wave action, and pulling open shells of bivalves.

Class Asteroidea	Class Echinoidea
Flattened star-shaped body	More or less spherical body
Tube feet in open ambulacral grooves	Tube feet penetrate pores in plates covering ambulacral grooves
Flexible body, since calcareous plates of skeleton not fused together	Calcareous plates fused to form rigid test
No jaw apparatus	Jaw apparatus present (Aristotle's lantern)
Carnivorous	Herbivorous
E.g. **Asterias rubens** (common starfish)	*E.g.* **Echinus esculentus** (sea-urchin)

Asterias rubens

External features (x ½)

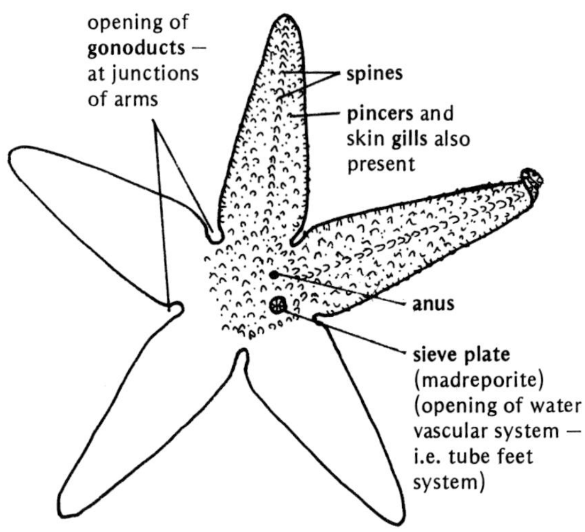

Upper (aboral) surface

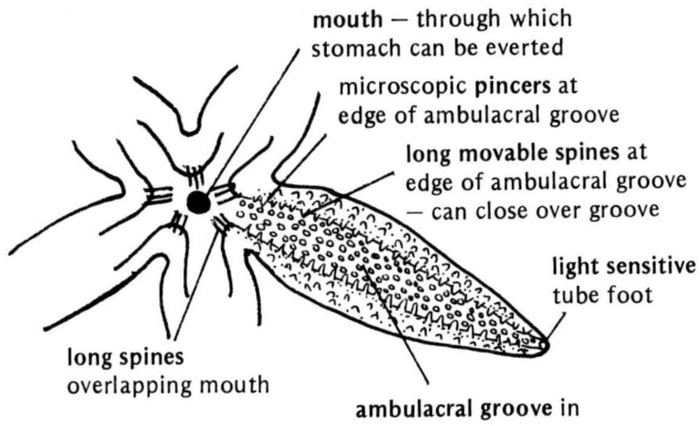

Lower (oral) surface

Notes

1. Occurs in coastal waters and between tide marks
2. Can move only slowly, but feeds on slow moving prey, mainly bivalves. Prey digested mainly externally; valves of shell pulled apart, and stomach of starfish inserted around contents.
3. Considerable powers of regeneration

Echinus esculentus

External features (x ⅔)

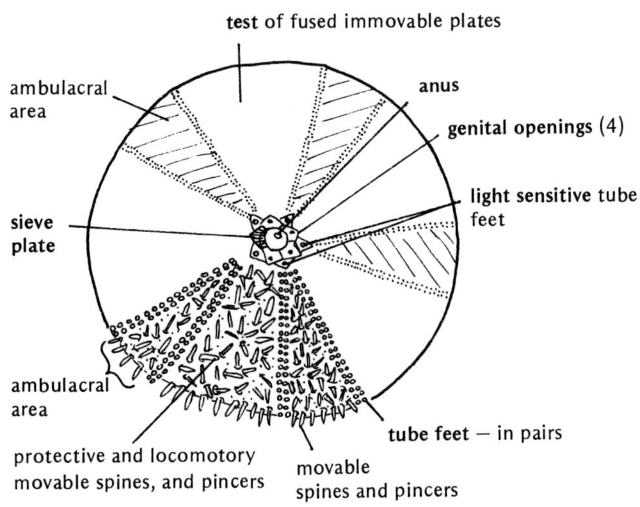

Upper (aboral) surface

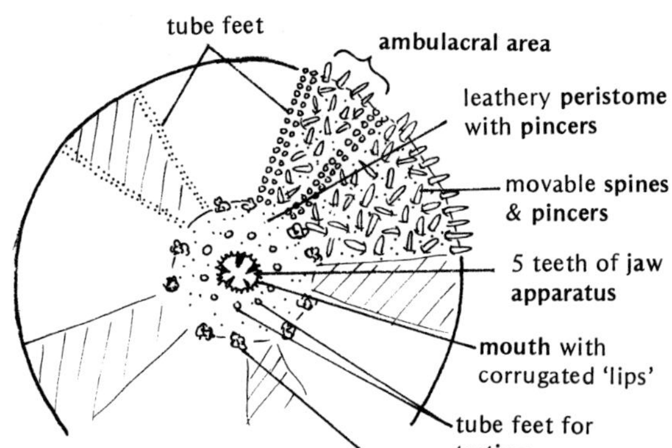

Lower (oral) surface

Notes

1. Common in the sub-littoral region of rocky coasts — sometimes found at low tide level.
2. Feeds mainly on algae, browsing rock encrusting vegetation.
3. The tube feet can be extended well beyond the spines — they enable the sea urchin to climb.

17. Phylum Chordata

General characteristics

Notochord present at some stage in life history – consists of a skeletal rod composed of highly vacuolated turgid cells enclosed in a tough connective tissue coat – extends along the length of the animal, dorsal to the gut and ventral to the CNS – provides a firm but flexible axis on which the segmental muscles can pull, permitting speedy locomotion through water.

CNS dorsal and tubular.

Paired visceral clefts perforate the lateral walls of the pharynx at some stage in life history – primitively provided filter feeding mechanism, but have acquired respiratory function in most aquatic forms (when known as branchial clefts or gill slits). In advanced terrestrial vertebrates the visceral clefts may never break through to the exterior, and are distinguishable only in the embryo.

Heart, when present, ventral – pumps blood forwards in the ventral aorta and backwards in the dorsal aorta(e) – these vessels are connected by paired 'arterial arches' in the pharyngeal region.

Metamerically segmented post-anal tail usually present.

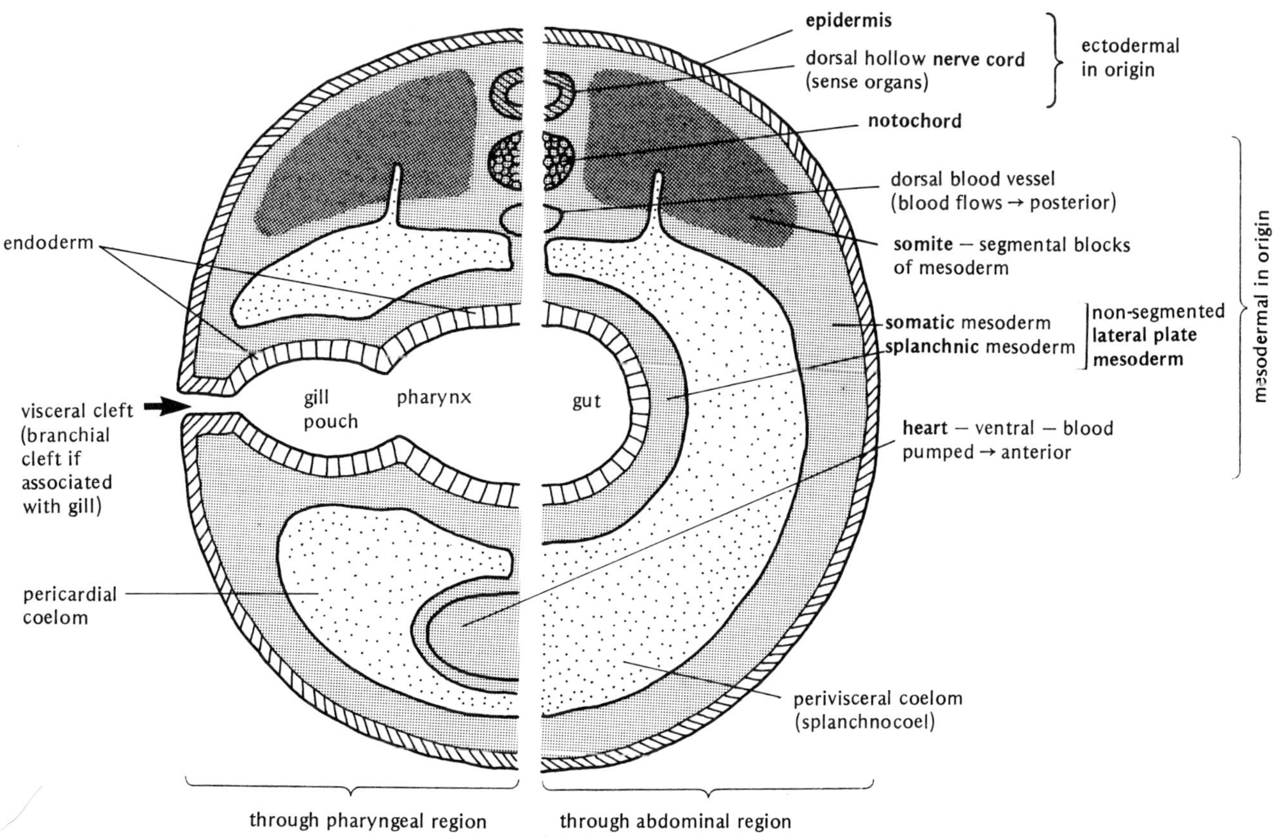

The chief structural features of a chordate as seen in transverse sections

Phylum Chordata

18. Sub-phylum Protochordata (Acrania)

General characteristics

Marine. Invertebrate. No definite heart. No kidneys.
Otherwise little in common, and sometimes each class is classified as a separate sub-phylum.

Class Hemichordata (acorn worms)	**Class Urochordata** (sea squirts)	**Class Cephalochordata** (lancelets)
Worm-like	Sac-like contractile body enclosed in test of substance similar to cellulose – two openings	Fish-like
Burrowing	Sessile – fixed to rocks/seaweed/ pier piles, etc – often colonial	Free-swimming, but spends most of the time buried in the sand with only anterior end protruding
Body divided into proboscis, collar and trunk – notochord confined to proboscis – terminal anus – not metamerically segmented	Adult degenerate – shows no chordate characteristics except gill slits – no coelom – not metamerically segmented	Full complement of chordate characteristics – metamerically segmented
Feeds by ingesting sand / mud	Ciliary feeding	Ciliary feeding
Sexes separate – external fertilization – larva very similar to larva of Echinoderms.	Hermaphrodite – external fertilization – free-swimming larva resembling tadpole and exhibiting **distinct chordate characteristics**	Sexes separate – planktonic larva
E.g. the following genera:		
Balanoglossus **Saccoglossus**	**Ascidia** **Ciona**	**Branchiostoma** (amphioxus)

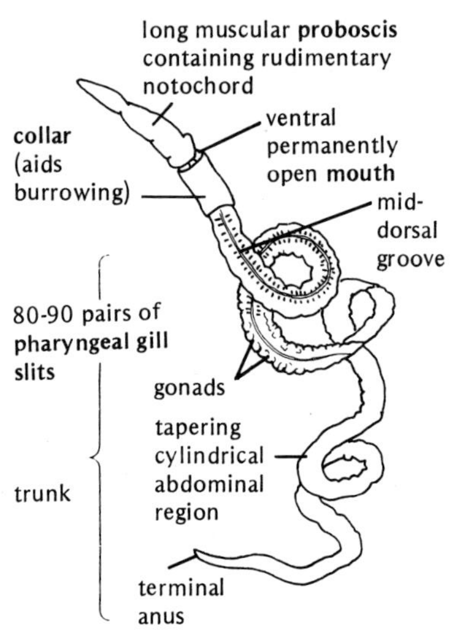

Saccoglossus cambrensis – external features (x 2)

Occurs in U-shaped burrow in sand or shell gravel containing organic debris – lower shore

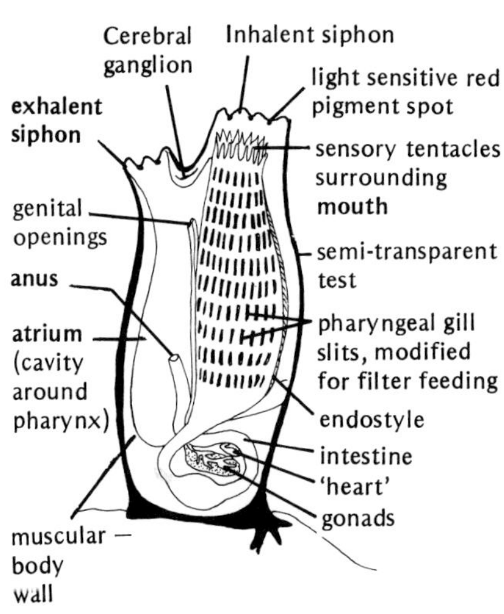

Ciona intestinalis – diagram to show structure (x ½)

Very common in shallow waters – particulary on buoys, piles etc – strains plankton from sea water

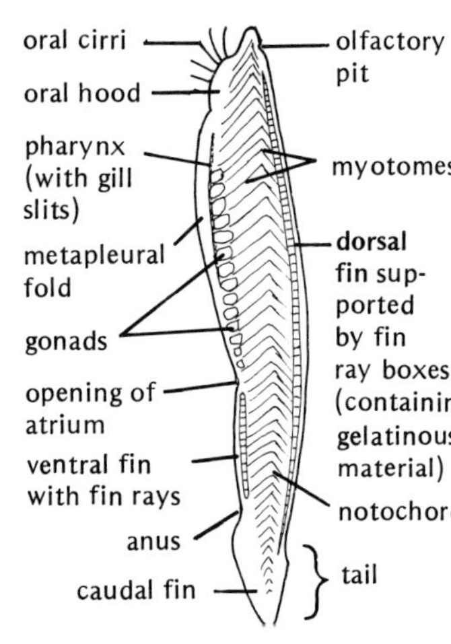

Branchiostoma lanceolatum – external features (x 1½)

Locally common in shallow seas – world wide distribution – feeds by straining plankton from sea water

CLASS CEPHALOCHORDATA

e.g. *Branchiostoma lanceolatum* (amphioxus/lancelet)

Diagram to show structure (Gonads omitted)

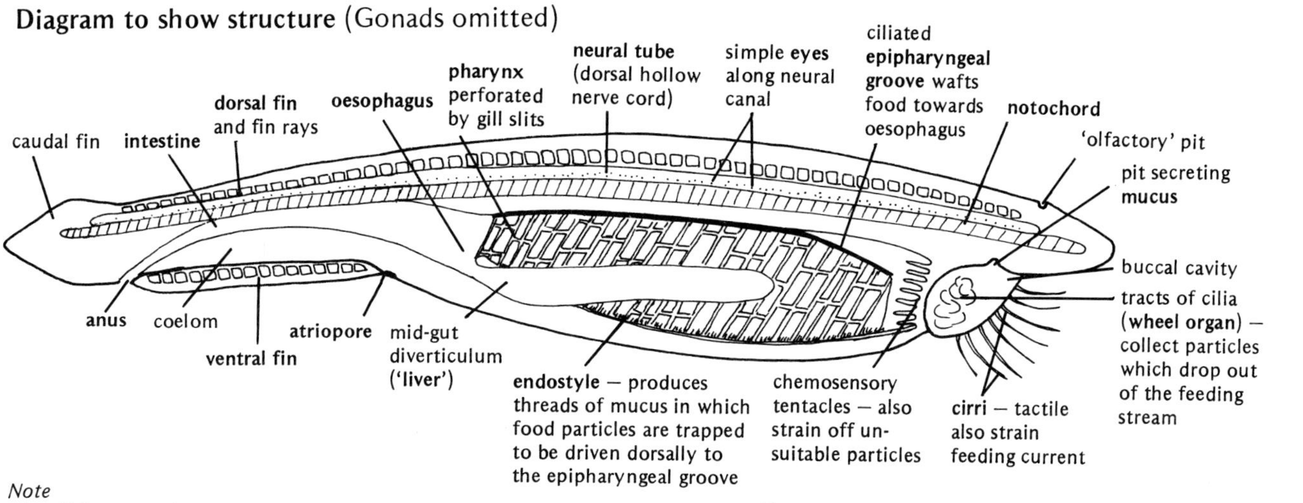

Note
The gill bars are of two types, primary gill bars (which contain a coelomic space) alternating with secondary gill bars, which develop later as downgrowths from the dorsal wall of each gill slit. Strengthening cross connections (synapticula) later develop

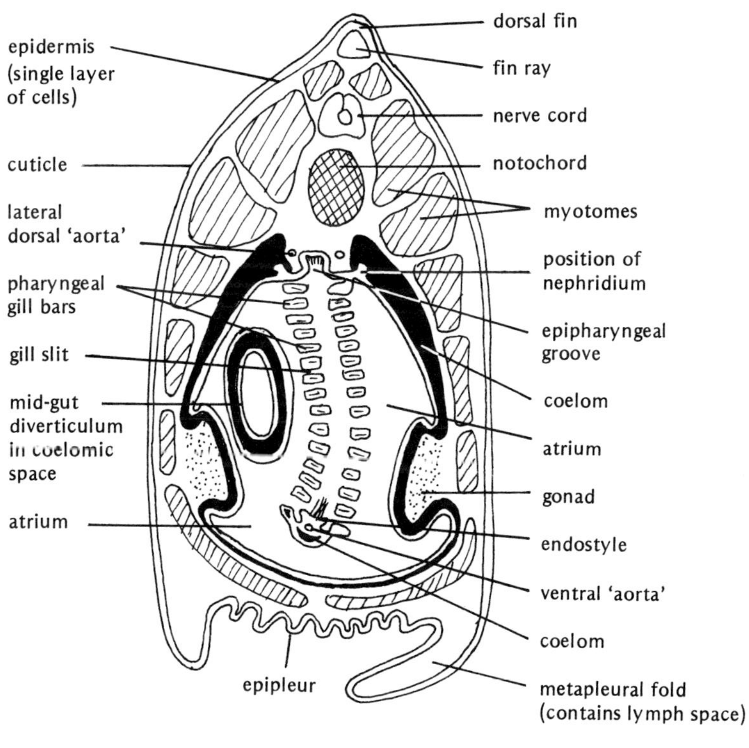

TS through pharyngeal region
(Diagrammatic, simplified)

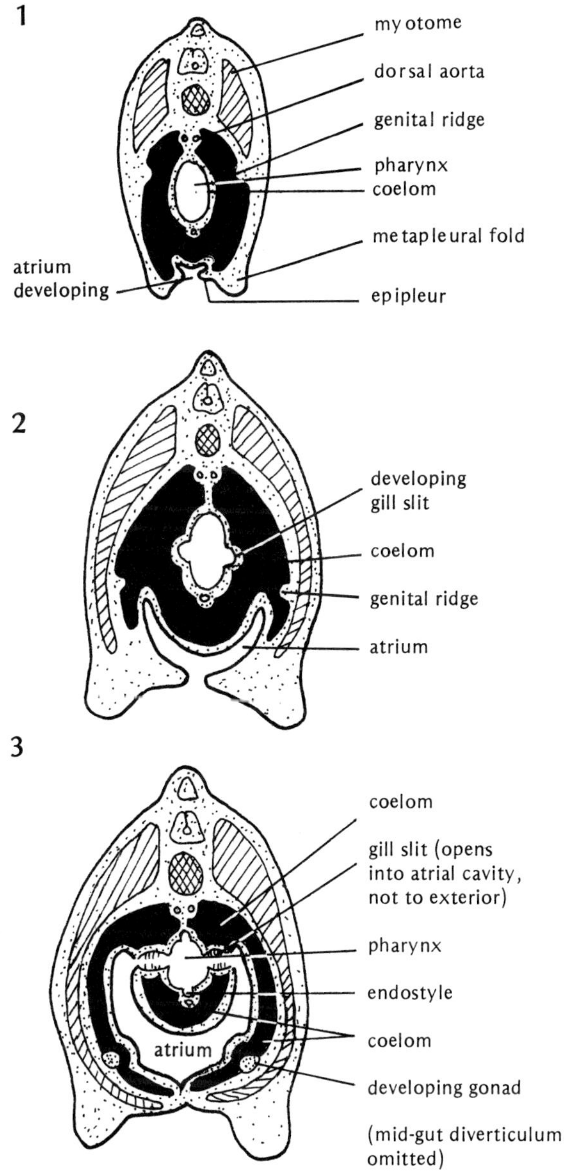

Diagrams to explain the origin of the atrium
(Highly idealised)

19. Sub-phylum Vertebrata (Craniata)

General characteristics

The notochord is wholly or partially replaced by a vertebral column in the adult. This is composed of a large number of small skeletal elements (vertebrae) extending the length of the body and surrounding the spinal cord. Each vertebra occupies an intersegmental position. The division into separate vertebrae allows greater freedom of movement without loss of strength

The head is clearly differentiated, the brain being enclosed in a cranium (brain box) and the sense organs (eyes, ears and olfactory organs) being well developed.

6 visceral clefts are typically present — they are not associated with feeding.

A very muscular heart is present, and blood circulates at high pressure.

No nephridia are present. The organs of nitrogenous excretion are kidneys which are mesodermal in origin, and which depend on high blood pressure.

The sexes are always separate, and asexual reproduction does not occur.

Most vertebrates have jaws and paired limbs (pectoral and pelvic), but the Agnatha do not.

CLASS AGNATHA (lampreys and hagfish)

Unjawed vertebrates — also known as *Cyclostomata*

General characteristics

Aquatic — eel-like — ectoparasitic (on true fish).

Rudimentary cartilaginous skeleton — vertebral column feebly developed, notochord persists in adult.

No jaws — mouth round and suctorial — tongue armed with rasping teeth.

No paired fins or girdles.

Numerous gills.

No scales.

Olfactory organ median, unpaired — only one or two semi-circular canals in ear — pineal body sensitive to light.

In lampreys the eggs hatch into free-living freshwater ammocoete larvae — ciliary feeding, with endostyle — live buried in mud.

Larval stage lasts up to 5 years.

e.g. *Petromyzon marinus* (sea lamprey)

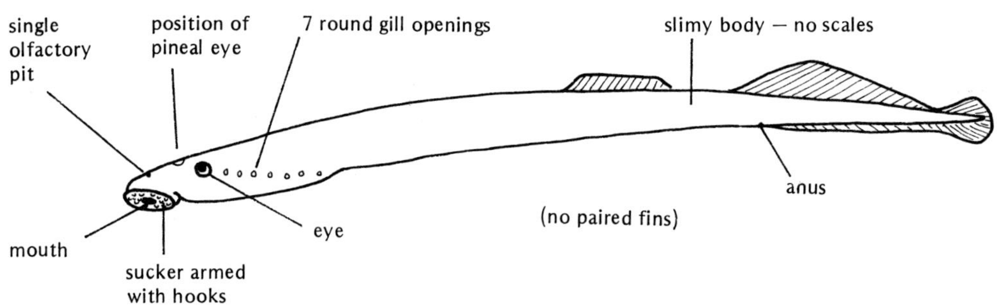

Notes

1. Grows up to 1 m long.
2. Migrates up-river from the sea to spawn in streams — then dies.

JAWED VERTEBRATES

Phylum chordata
Sub-phylum
Vertebrata

The basic organisation of the head end of a vertebrate

Note
Skull = cranium + visceral arch skeleton

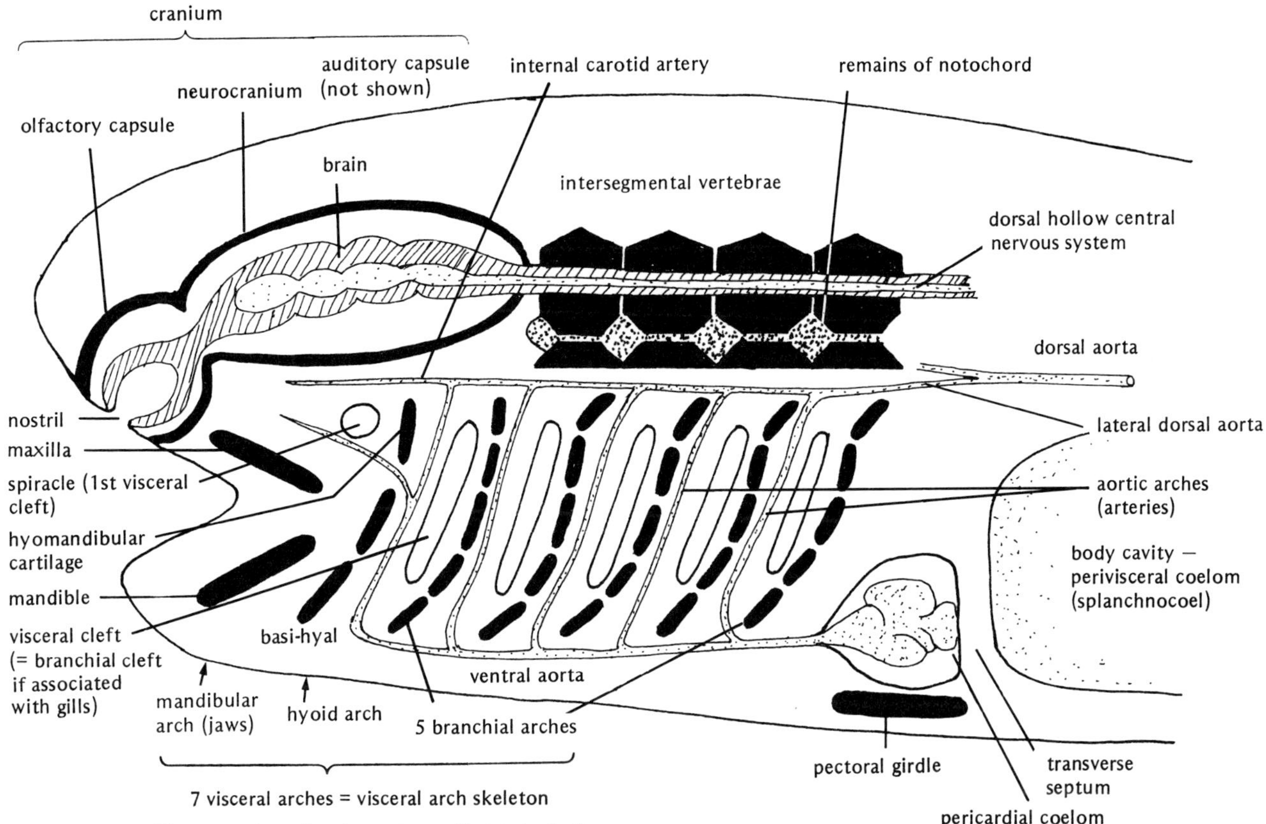

The term visceral arch can be applied to both the tissues between adjacent visceral clefts and to merely the skeleton supporting this tissue

The basic segmentation of the vertebrate body
(i.e. portion of **embryo** — highly diagrammatic)

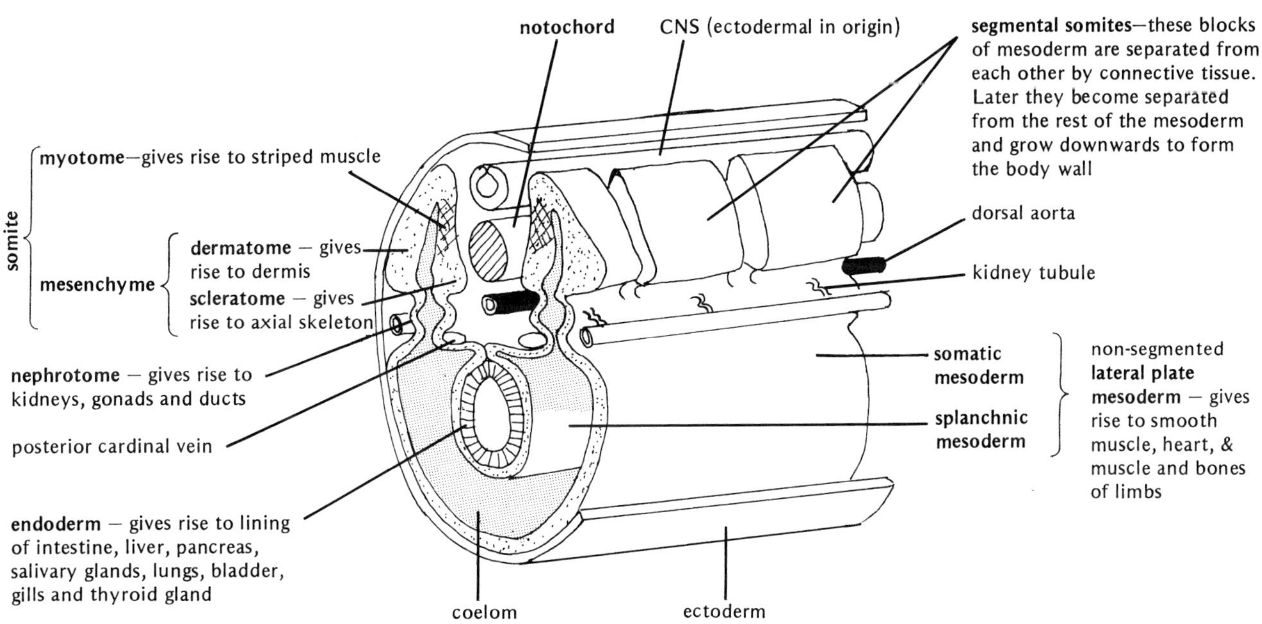

Phylum chordata
Sub-phylum
Vertebrata

FISHES

Fishes are not now included in a single class. Present-day fish fall into two classes (Elasmobranchs and bony fish), but they have several features in common

General characteristics

They are vertebrates which are completely adapted to an aquatic life.

The body is streamlined with a muscular propulsive tail.

Paired pectoral and pelvic *fins* are always present, also median unpaired fins which maintain stability.

The fins possess an endoskeleton, muscles, and fin rays – they are not merely folds of skin.

The respiratory organs are gills, borne on branchial arches. Accessory breathing organ(s) may be present.

The circulation is single.

Prominent lateral line system is present.

Exoskeleton of scales.

No internal nares, except in a few lung-fish.

Mucus-secreting glands in skin – slimy body minimises friction.

Ectothermic i.e. poikilothermic/cold blooded.

Class Elasmobranchii (Chondrichthyes)	**Class Osteichthyes (Bony fishes)**
Cartilaginous endoskeleton, flexible body	Bony skeleton – cartilaginous skeleton only in embryo
Exoskeleton of placoid scales which pierce skin – they are continuously replaced	Exoskeleton of bony scales, below the surface of the skin – they grow with the fish
Gill slits not covered by operculum	Operculum present
Spiracle present	No spiraclar opening
Well developed sense of smell – poor vision	Well developed colour vision
Mouth ventral (even in skates and rays)	Mouth typically terminal
No air bladder – body heavier than water	Air bladder present – usually hydrostatic – may function as lung
Broad horizontal paired fins, act as elevators.	Fins variable – paired fins do not act as elevators – pelvic fins tend to be more anterior in position
Heterocercal tail	Tail fin usually symmetrical
Primitive jaw suspension – usually hyostylic	Variable jaw suspension
Single pair of cartilages in upper and lower jaws	Numerous jaw bones
Spiral valve in intestine. Cloaca	Anus and urinogenital apertures usually separate
Internal fertilization – eggs enclosed in horny capsule.	External fertilization
E.g. **Sharks, skates, rays**	E.g. **Most common fish, and lung-fish**

Notes

1. Fish possibly originated in fresh water, their tissues are hypotonic to sea water
2. The body shows counter-coloration, i.e. dark dorsally and lighter ventrally. This renders the fish less conspicuous.
3. There is no external sign of the ears – only the inner ear is necessary in aquatic vertebrates.
4. The fins are not usually propulsive, apart from the caudal fin. The 'paired fins' are concerned with stopping, steering, and raising and lowering the head.

CLASS ELASMOBRANCHII

Phylum chordata
Sub-phylum
Vertebrata

e.g. *Scyliorhinus caniculus* (dogfish)

External features (x $\frac{1}{5}$)

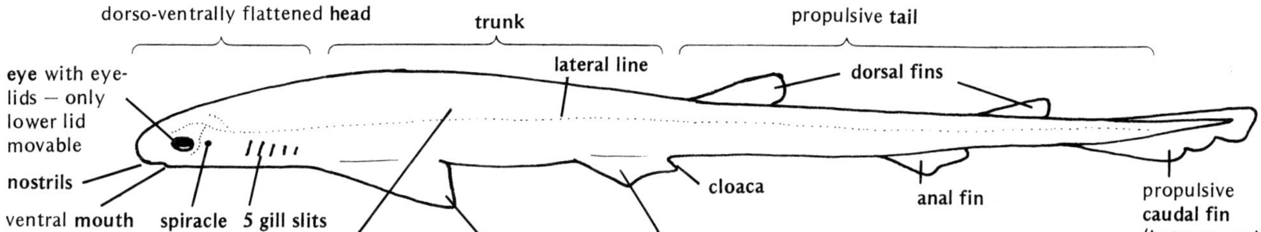

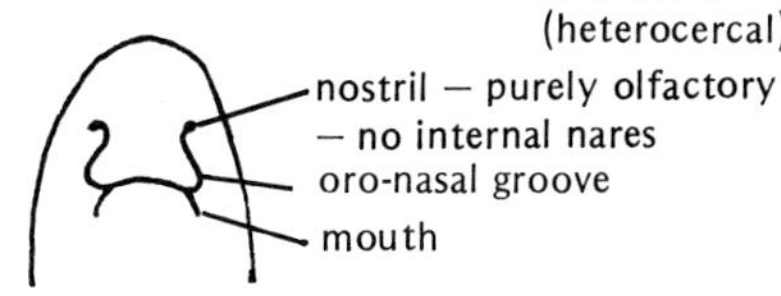

Ventral view of head

Notes

1. Very common fish in coastal waters
2. Shape of body important with regard to buoyancy.
3. Carnivorous scavenger – preys on relatively slow moving animals of the sea floor, e.g. molluscs, crustaceans, worms, etc. Relies mainly on sense of smell and sensitivity to vibrations to locate prey. Eyes adapted to near vision in poor light. Eyelids protect the eyes when the fish is seeking prey in mud (eyelids are unusual in fish).
4. Cannot swim backwards (therefore there is no disadvantage in the placoid scales pointing backwards) – neither can it stop abruptly.
5. Sexual dimorphism – male possesses claspers. Fertilization is internal, and only a few large yolky eggs are produced, each protected in a horny egg case. Direct development.

CLASS OSTEICHTHYES

e.g. *Scomber scombrus* (mackerel)

External features (x ½)

**Note*
In most boney fish including the mackerel each nostril has 2 openings to the exterior, allowing water to circulate efficiently within the olfactory organ.

Whole body provides propulsion

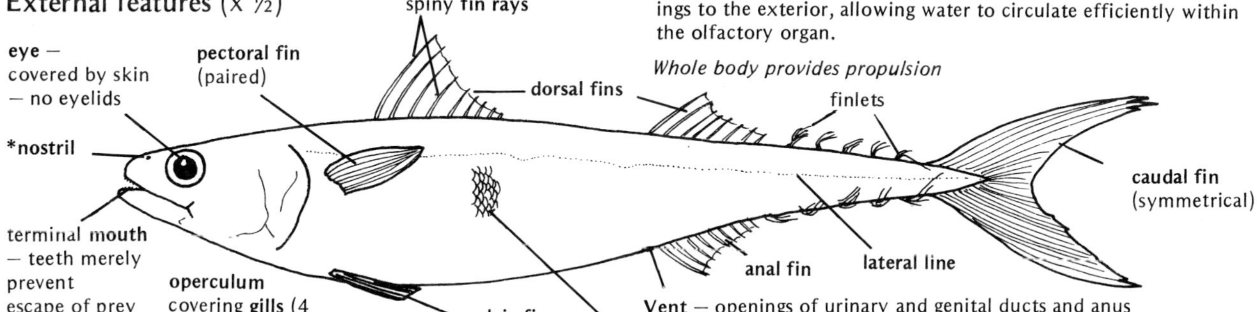

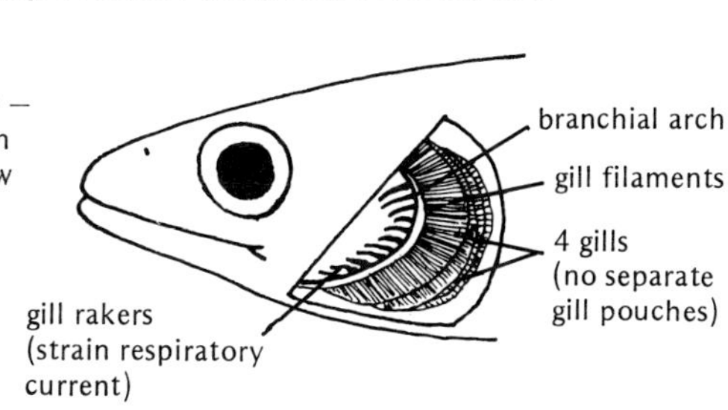

View of gills
(operculum removed)

Notes

1. A voracious and predatory fish which can be found almost anywhere around the coast of the British Isles in the summer months. Mackerel winter in shoals near the bottom of the sea, off the coast of Ireland and in the northern North Sea – rise from the deep in spring, and migrate to coastal waters where they feed on plankton and then spawn. After spawning the shoals break up into smaller groups, and the fish then forage inshore at the surface, feeding on small fish and fry of all species, including their own.
2. Laterally flattened body – this is the most efficient shape when locomotion is achieved by lateral undulations of the body.
3. No swim bladder (this is unusual in bony fish). So although the fish sinks when it stops swimming, it can rise or swim to great depths very rapidly to avoid predators.

20. Tetrapods (amphibia, reptiles, birds & mammals)

Terrestrial vertebrates.

General characteristics

Bony skeleton

2 pairs of pentadactyl limbs.

The pelvic girdle fused to the vertebral column.

Lungs and a double circulation of some sort.

Middle ear present, for the transmission of aerial sound waves to the fluid of the inner ear.

Rounded cornea to eyes – eyelids present, and lachrymal glands (c.f. windscreen wipers and washers)

Generalised tetrapod skeleton

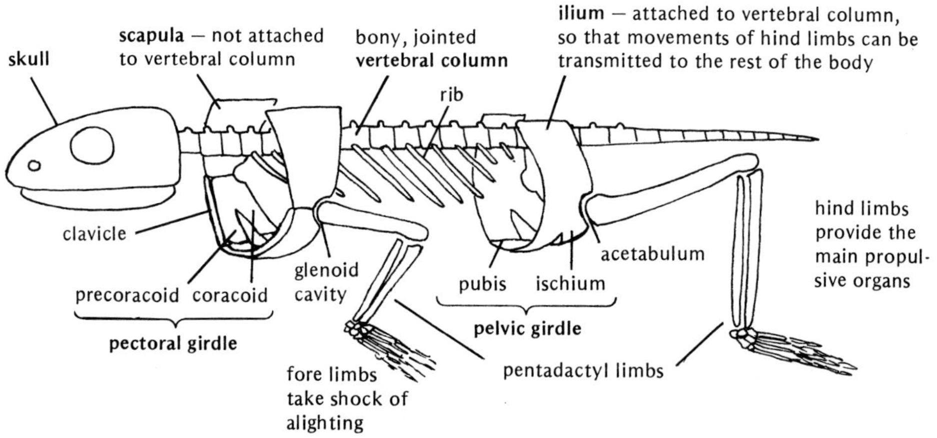

The skull, ribs and vertebral column provide the axial skeleton.
The girdles and the skeleton of the limbs provide the appendicular skeleton.

The pentadactyl limb

The skeletons of the limbs of all tetrapods show a remarkable constancy of pattern despite differences in locomotion. Thus the following generalised plans are applicable to the limb skeletons of amphibia, birds, reptiles and mammals, including dolphins, bats, moles, etc. Modifications involving the loss or fusion of bones occur in different species, of course.

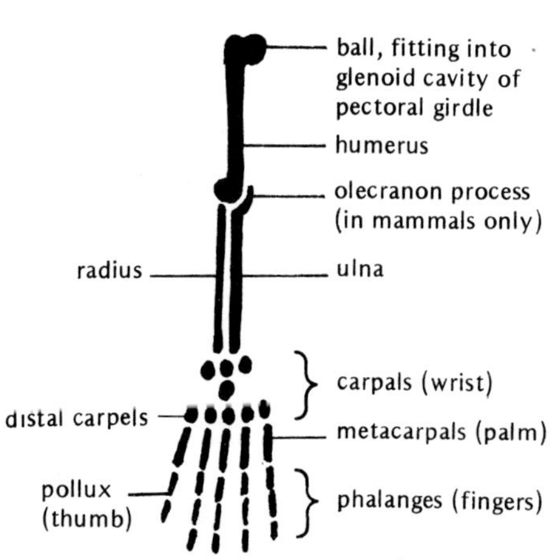

L. fore limb skeleton

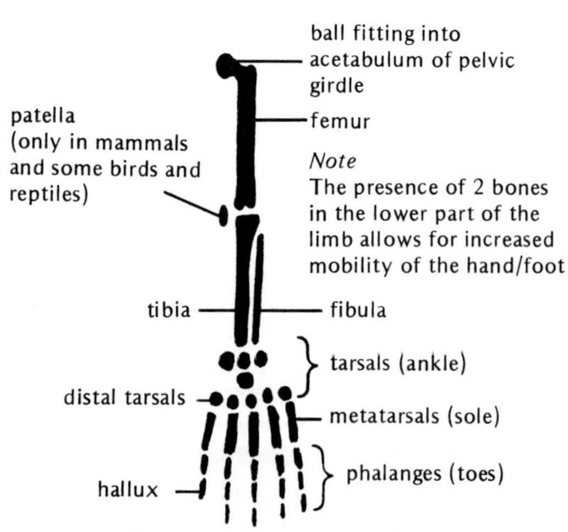

L. hind limb skeleton

The earliest tetrapods probably moved rather like newts and lizards, where rapid locomotion involves serial contraction of muscles along the body (as in fish) and the body is hardly lifted from the ground except during slow and deliberate movements.

The re-orientation of the limbs during evolution to occupy a position beneath the body, (where they can act as vertical *levers* taking the weight of the body more along their axes), is much more efficient, and reduces the necessity for many of the muscles previously required to keep the body clear of the ground.

Primitive orientation of limbs

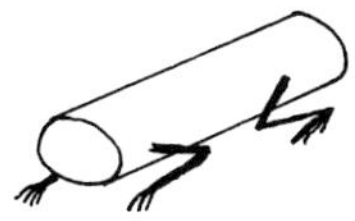

Limbs flexed and rotated from primitive condition

CLASS AMPHIBIA (frogs, newts, salamanders, etc.)

General characteristics

Ectothermic (poikilothermic/cold-blooded)

Smooth moist skin – not waterproof – glandular – provides respiratory surface.

Teeth uniform (homodont).

10 cranial nerves.

No external ear – single ossicle (Eustachian tube and middle ear derived from the hyoid cleft, i.e. spiracle).

Urea and ammonia excreted – mesonephric kidney.

Cloaca present.

Eggs laid, unprotected by shell – fertilization external – eggs laid in fresh water.

Indirect development – larva aquatic, possessing gills – no parental care.

Semi-terrestrial; dependent on fresh water for development.

ORDER URODELA (Newts and Salamanders)

General characteristics

Elongate body with tail.

Neck present – head can be turned.

Limbs all short and of equal length.

No ear drums.

Voiceless.

Well developed sense of smell.

Adult and larva differ little from each other.

Restricted to damp habitats.

ORDER ANURA (Frogs and Toads)

General characteristics

Short body.

No tail.

No neck – head cannot be turned.

Hind limbs much longer and more muscular than front, with webbed feet.

Ear drums present.

Males able to croak.

Sight dominant sense.

Long sticky mobile tongue, attached anteriorly and used for catching flying prey.

Phylum chordata
Sub-phylum
Vertebrata
Class Amphibia

ORDER URODELA

e.g. *Triturus vulgaris* (smooth newt)

External features (x 1½)

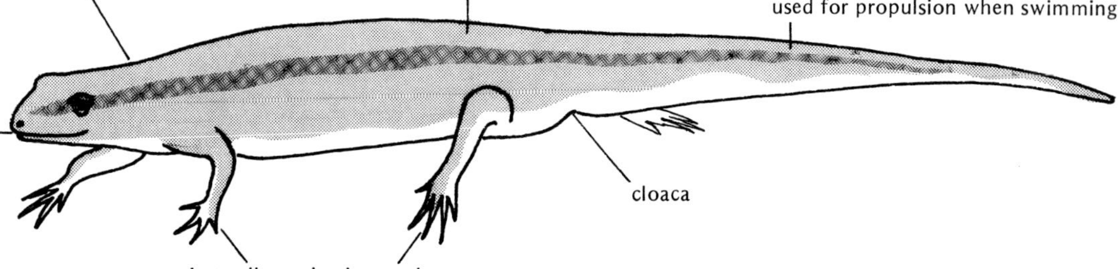

Notes

1. Widely distributed. Found near ponds but seldom enter water except in the breeding season.
2. Hide in crevices and long grass during daylight.
3. Voracious nocturnal feeders, eating worms, slugs, snails and insects. Can feed in water, on small crustacea, aquatic larvae, molluscs and tadpoles.
4. Hibernate during winter under stones, etc.
5. Adult males are slightly smaller than females. They are voiceless.
6. The breeding season is in spring. After emergence from hibernation the mature animals make their way to a pond. They become brightly coloured and the male develops a high undulating crest extending along the length of the body. A complex courtship process takes place during which the male releases sperm in a gelatinous capsule (spermatophore). The female lowers her body over this and presses it into her cloaca. The spermatozoa are stored in the roof of the cloaca and the eggs are fertilized as they are laid, singly, over a period of weeks. Each jelly-coated egg is attached to a leaf. The tadpole larva retains feathery external gills until metamorphosis is completed, at the end of summer or in the following spring. (The fore-limbs appear before the hind.) The miniature newts leave the water and do not return until sexually mature, about 3 years later.

ORDER ANURA

Phylum chordata
Sub-phylum
Vertebrata
Class Amphibia

e.g. *Rana temporaria* (common frog)

External features (x 1½)

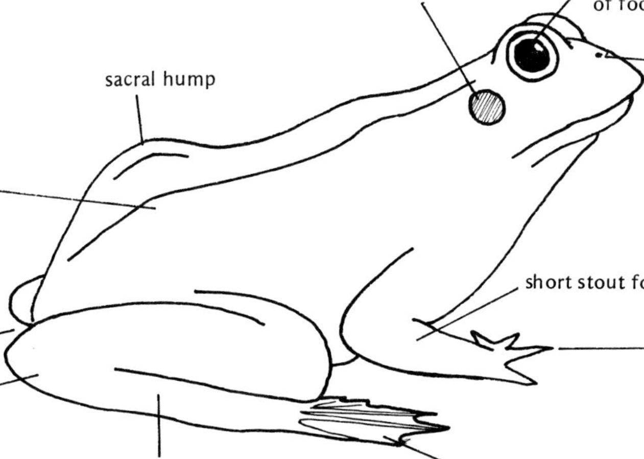

Note
Streamlined body, no neck — shape adapted for swimming. Loss of tail gives greater mobility on land

Notes

1. Found in damp places, feeding on moving invertebrates (earthworms, slugs, grubs and flying insects). The capture of flying prey is effected by means of the sticky extensible tongue.
2. The skin provides the main respiratory surface. Lungs are used only spasmodically, usually when the frog is active on land.
3. The breeding season is in early spring when the frogs emerge from hibernation. The males croak to attract the females to a suitable spawning pond. Each male mounts a female clasping her firmly by means of nuptial pads which develop seasonally on his fore limbs. Seminal fluid is poured over the eggs as they emerge. On contact with water the albumen coating the eggs swells, and the eggs are held together in large floating masses. The eggs hatch after about 14 days. The larva is a fully aquatic tadpole which is at first herbivorous. Metamorphosis occurs gradually, the miniature frogs leaving the water in late summer/autumn. Sexual maturity is reached 3–4 years later. No parental care is given at any stage. The mortality rate is very high; a great number of eggs are produced to compensate for this.

Phylum chordata
Sub-phylum
Vertebrata

CLASS REPTILIA (lizards, crocodiles, snakes, turtles, etc.)

General characteristics

Ectothermic – no hair or feathers.

Skin waterproof, dry, not glandular; horny epidermal scales present.

Lungs provide the sole respiratory organs (but there are no regular rhythmical breathing movements)

Vertebral column usually long and flexible.

Distinct neck usually present.

Teeth uniform (homodont).

12 cranial nerves.

No pinna to ear – external auditory meatus sometimes present – single ossicle – lagena often well developed. Sense of smell not well developed, since breathing not regular – Jacobson's organ present in snakes and lizards for 'tasting' air sampled by flicking tongue. (Snakes are stone deaf)

Uric acid excreted – metanephric kidney.

Cloaca present.

Large shelled eggs laid – yolk laden – internal fertilization – direct development – little parental care.

Completely adapted to life on dry land, both as adults and during development.

e.g. *Lacerta vivipara* (common lizard)

External features (x 1)

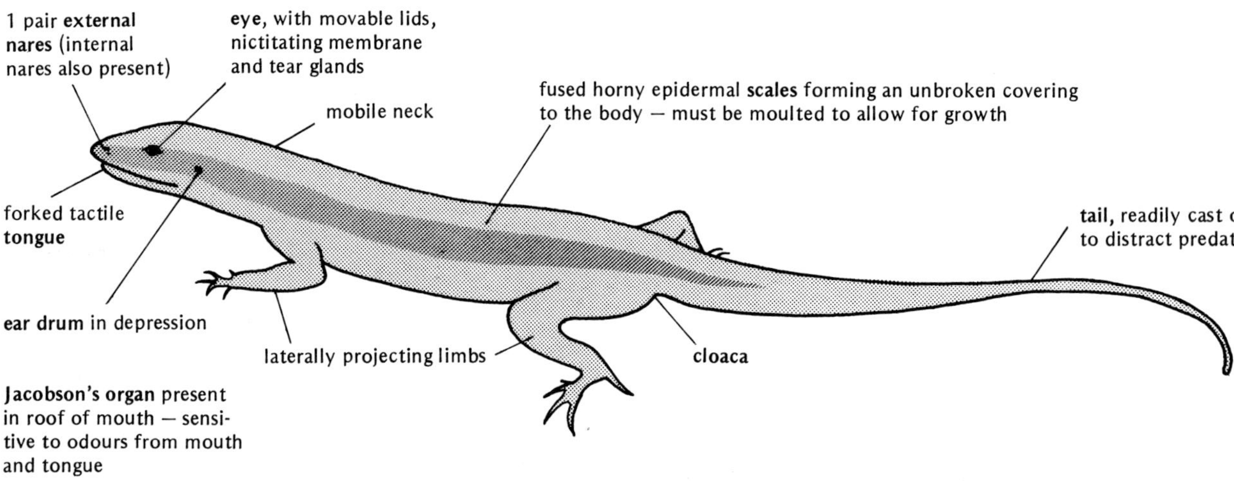

Notes

1. Common in dry sandy heathlands; lives in a hole in a wall or bank.
2. Feeds on insects, worms, snails, etc (carnivorous)
3. Very active, agile and fast moving, once warmed up by the sun.
4. Diurnal.
5. Preyed upon by snakes, weasels, some birds, and shrews – main defence is rapid retreat, often casting off tail to create a distraction
6. Hibernates in winter.
7. Males fight for possession of females – eggs retained in mother's body, being laid in August, at the moment of hatching.

e.g. *Natrix natrix* (grass snake)

Phylum chordata
Sub-phylum
Vertebrata
Class Reptilia

External features (x ¾)

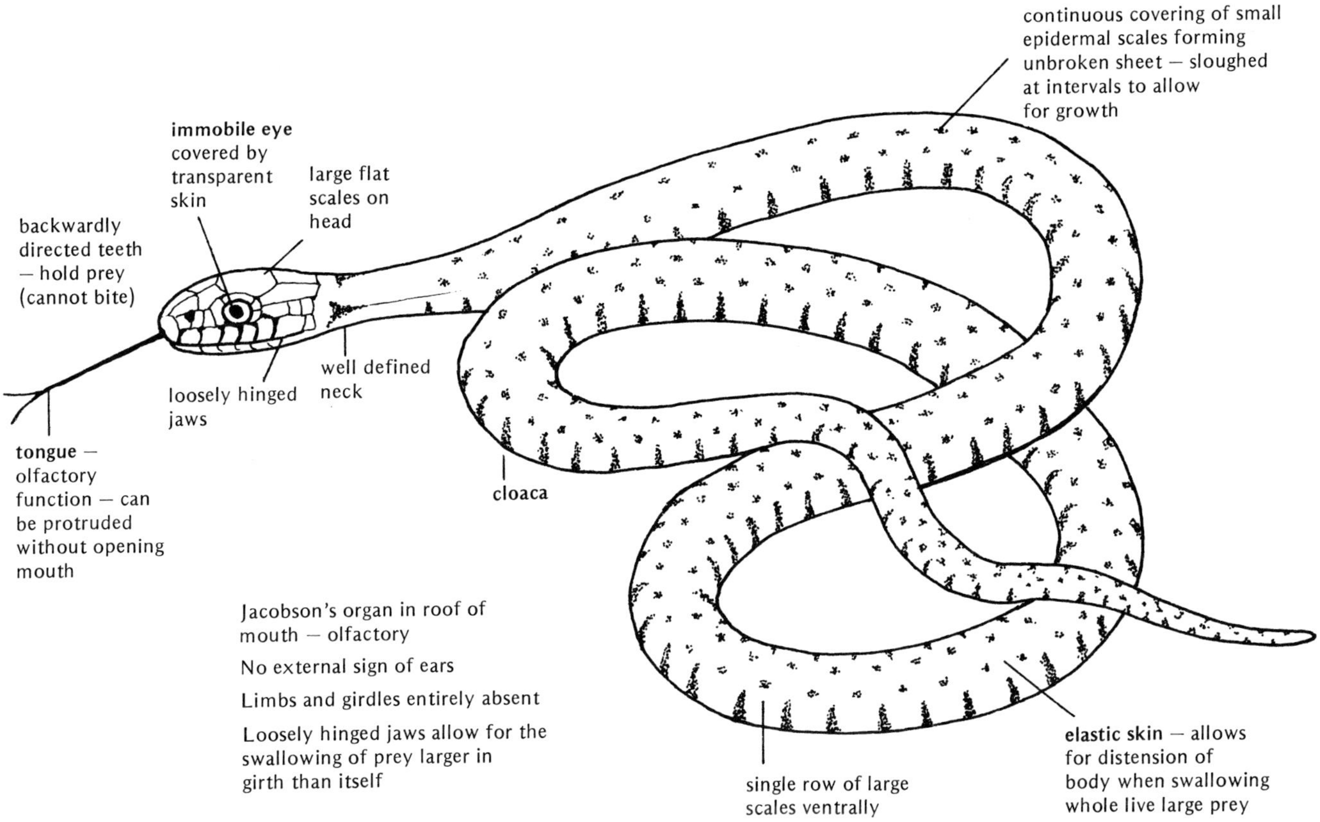

Notes

1. Found on rough commons with long grass and bushes, and in deciduous woods with large clearings – usually in the vicinity of a pond, stream or river, and more commonly in the south of England.

2. Non-venomous. Feeds on cold-blooded prey, mainly amphibia and fish, which are located by sight, smell and vibrations. The grass snake has no means of killing or subduing prey, and can therefore capture only cold-blooded animals which do not struggle or retaliate. But fledglings and freshly dead mice and birds can be eaten. Fish and tadpoles can be captured in water, but fish are eaten on land, as the snake must breathe air. Prey are swallowed alive. Only one or two meals are needed a week and only when the snake is active.

3. Moves effortlessly by means of rippling muscles. Can mount vertical surfaces. Throws body into a series of loops (side winding) during fast movement and when swimming.

4. Diurnal – actively hunts only in summer, after basking in the sun to get warmed up.

5. Mating occurs in late spring, after basking and courtship. 30–40 eggs with tough leathery skins are laid in a warm damp place (e.g. under piles of decaying matter) in July. They are coated in mucus which sticks them together as it dries. The eggs hatch about 8 weeks later, the infant snake (18cm long) piercing the shell with an egg tooth, which is soon shed. Sexual maturity is reached after 3–4 years when about 60–70cm long.

6. Hibernates from October to April, buried in loose soil among roots, or in a disused rabbit's burrow, etc.
The elongate body, loss of limbs and reduction of sense organs indicate that snakes have probably evolved from burrowing ancestors. The eyes have re-evolved, but remain covered by skin (transparent).

Phylum chordata
Sub-phylum
Vertebrata

CLASS AVES (birds — vertebrates adapted for flight)

General characteristics

Endothermic (homoiothermic/warm-blooded).

Skin waterproof — non-glandular, apart from preening gland at base of tail. Feet scaly, rest of body covered by epidermal feathers, composed of keratin, enabling flight and maintenance of a high body temperature.

Lungs provide the sole respiratory organs — rhythmical breathing movements occur — highly efficient respiratory system.

Bones light and strong — many contain air spaces into which project air sacs extending from the lungs.

Pectoral limbs modified to form wings, with large surface area. Large keeled sternum for attachment of pectoral muscles — strong pectoral girdle.

Bi-pedal method of walking — hind limbs large and strong with considerable fusion of bones. Large pelvic girdle, rigidly fused to synsacrum (fused lumbar and sacral vertebrae) allowing weight of body to be carried entirely on hind limbs when not flying. Digitigrade.

No teeth — jaws ensheathed in horny beak.

12 cranial nerves.

No pinna to ear — external auditory meatus present — cochlea present — single ossicle.

Acute sense of sight — colour vision — extension visual field — perception of fine detail over visual field all parts of which are in focus at once.

Uric acid excreted — metanephric kidney.

Cloaca present.

Internal fertilization — large, yolk-laden eggs laid, enclosed in calcareous shell.

Direct development — nest constructed — considerable parental care, involving both parents.

Strong instinctive behaviour.

e.g. *Fringilla coelebs* (chaffinch)

External features (x 1)

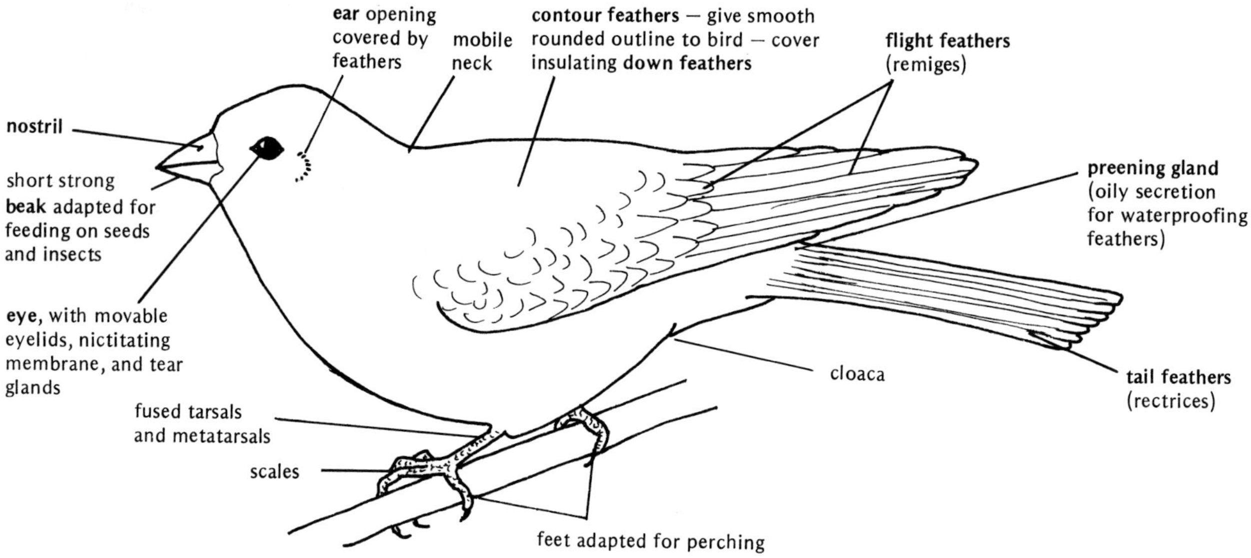

Notes

1. One of the most common birds in Britain.
2. Nomadic in winter — hunts for food in flocks — males leave flock in Spring to acquire a territory and then a mate.
3. Nest built by hen, in hedge, shrub or small tree — 4-6 eggs laid in May — incubated by hen after full clutch has been laid — fed by cock during this period.
4. Chicks hatch after 11-13 days — altricial (i.e. naked, blind and helpless) — must be brooded until their feathers develop sufficiently — then both parents can leave nest to hunt for food (mainly caterpillars) — chicks leave nest after 10 days, as fledglings which can at first neither fly well nor feed themselves adequately — must be tended by parents for some time before becoming fully independent. A second brood is usually reared later in the year.

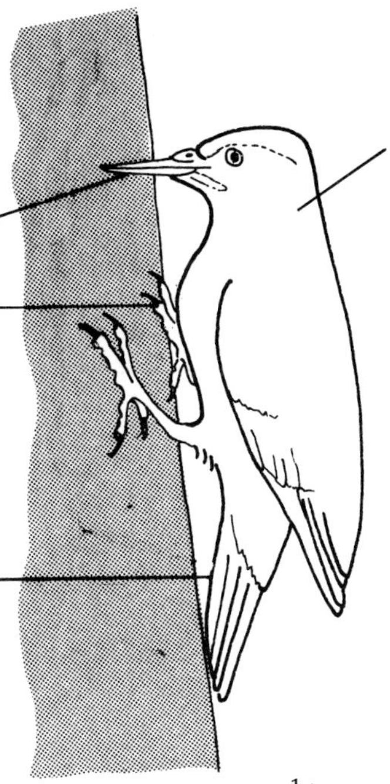

Picus viridis (x 1/5)

Picus viridis (green woodpecker)

Woodpeckers are tree climbing, hole-drilling birds. The green woodpecker is found in open woodlands and parks, seeking wood-boring larvae and insects in infected trees. It alights on the trunk and works upwards, tapping and drilling the bark with its powerful beak and 'winkling' out prey by means of its very long sticky tongue. It will also eat insects and worms from the ground; it is particularly fond of ants.

Tyto alba (barn owl)

The barn owl is a nocturnal bird of prey, depending mainly on its acute senses of hearing to locate prey. Its flight is silent due to specialisation of the wing feathers. It relys on taking its victims by surprise, hunting mainly small mammals (rats, mice, voles and shrews) but also small birds and insects. Prey is swallowed whole, the undigested residue being later regurgitated in the form of pellets.

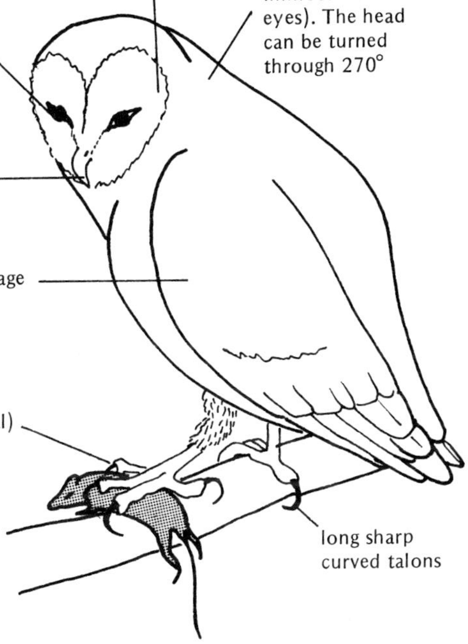

Tyto alba (x 1/5)

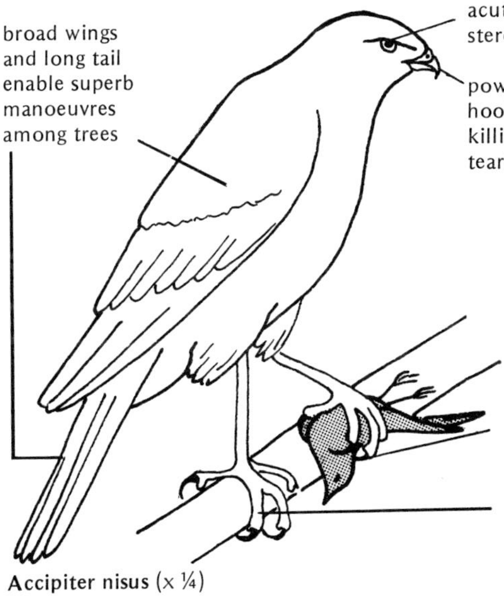

Accipiter nisus (x ¼)

Accipiter nisus (sparrowhawk)

The sparrowhawk is a diurnal bird of prey. It flies low, along hedgerows and in woods, hunting small birds, mice, frogs and insects. It can thread its way amongst trees and can take birds on the wing. The quarry is eaten on the ground, the sparrowhawk standing on it and ripping it apart. Undigested residue is later thrown up as pellets.

Phylum chordata

Sub-phylum

Vertebrata

Class Aves

Anus platyrhynchos (mallard duck)

The mallard duck is a swimming bird found in marshes, estuaries, sheltered coasts and all types of fresh water. It feeds, mainly at night, by drawing water through its beak and straining out food by means of the toothed edges of the lower jaw. It eats leaves, seeds, molluscs, larvae, insects, frogs and fishes from the water. It also grazes on vegetation and eats berries on land, and dabbles in mud and puddles.

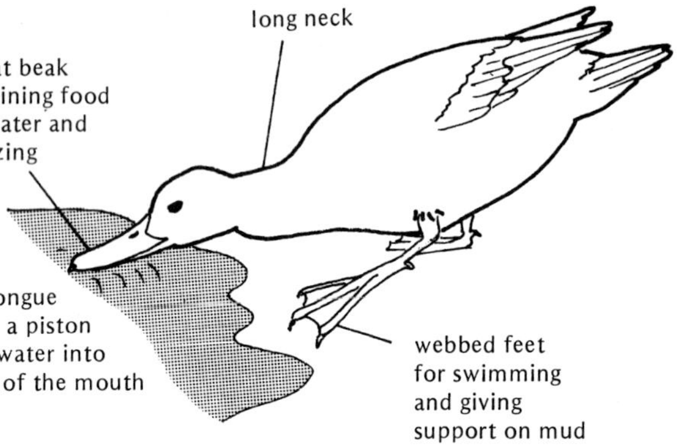

Anas platyrhynchos (x 1/8)

Ardea cinerea (heron)

The heron is a wading bird; it is adapted for walking in deep water to catch prey. It feeds mainly on fish and frogs, but can also catch young wildfowl, mice, rats, water voles, beetles, etc. on land. It may stand motionless waiting for quarry, at which it will stab with great accuracy and speed, or it may stalk prey. It is most active at dawn and dusk.

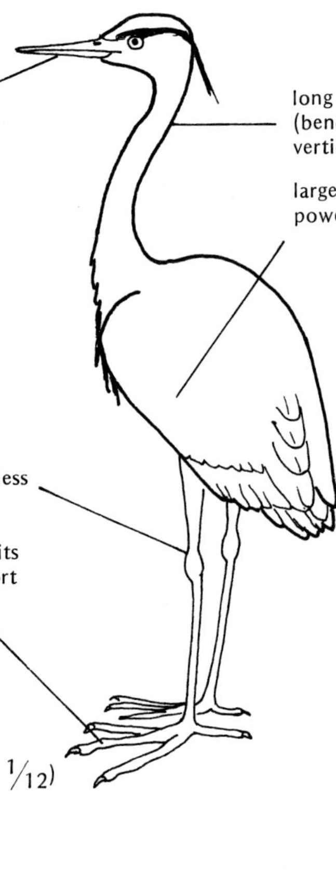

Ardea cinerea (x 1/12)

Troglodytes troglodytes (wren)

The wren is a perching song bird. It is small and inconspicuous and incessantly active as it hunts insects and spiders etc. in the depths of creepers, bushes and undergrowth.

Troglodytes troglodytes (x ¾)

CLASS MAMMALIA

General characteristics

Endothermic (warm blooded).

Waterproof hairy skin, rich in glands.

Lungs provide sole respiratory organs – regular rhythmical breathing movements – diaphragm present, greatly increasing the efficiency of the respiratory system.

Epiphyses to bones, allowing growth over a long period of time without disturbing the joints.

False palate present – i.e. roof of mouth formed by bony shelves projecting inwards from bones of the upper jaw – the true palate. The internal nares therefore open far back in the pharynx and food can be chewed in the mouth while breathing continues.

Single pair of bones forming the lower jaw (R and L dentary).

Heterodont – teeth of different types modified according to diet.

12 cranial nerves.

3 ear ossicles (2 evolved from bones of the lower jaw) – pinna to outer ear – cochlea present – dominant senses are usually hearing and smell; colour vision probably only in Primates. But all senses are well developed.

Urea excreted – metanephric kidney.

Separate anus and urino-genital aperture, except in monotremes.

Internal fertilization – viviparous, except for monotremes. Young nourished after birth by milk secreted from mammary glands (modified sebaceous glands except in monotremes) – considerable parental care.

Brain with large cerebral hemispheres and large complicated cerebellum.

Active, exploratory and inquisitive – capable of showing ingenuity and enterprise.

SUB-CLASS MONOTREMATA (egg-laying mammals – duck-billed platypus, spiny ant-eater, etc.)

Found only in Australia.

General characteristics

Shelled eggs laid carried in temporary pouch by ant-eater – laid in nest in burrow by platypus.

No true teeth – 'beak' present, covered by skin.

Milk secreted by specialised sweat glands – no teats – discharged into fold of abdominal skin.

Cloaca present.

Body temperature lower and more variable than that of other mammals.

Extremely primitive mammals.

SUB-CLASS MARSUPIALIA (pouched mammals)

The dominant mammals of the Mesozoic with world-wide distribution. Now confined mainly to Australasia (there are a few species in South America, and a single species in N. America). They became geographically isolated in Australia before the placental mammals had invaded this region. Their adaptive radiation parallels that of the placental mammals which replaced them in the rest of the world. Thus there are cursorial kangaroos, arboreal opossums and koalas, burrowing wombats, mice and moles, carnivorous Tasmanian wolves, native cats, etc.

General characteristics

No placenta.

Young born at a very early stage of development – make their own way to the pouch (marsupium), which contains the teats of the mammary glands – the young become attached to the teats until their development is completed.

No milk dentition – i.e. only one set of permanent teeth.

Oviducts completely separate – two distinct uteri.

Phylum Chordata
Sub-phylum Vertebrata
Class Mammalia

SUB-CLASS EUTHERIA (placental mammals)

General characteristics

Young born at a very advanced stage of development, since nourished by means of an allantoic placenta enabling prolonged gestation.

Two sets of teeth – milk dentition replaced by permanent teeth.

Single uterus, though it may be divided into two 'horns'.

e.g. *Oryctolagus cuniculus* (wild rabbit)

External features (x ⅕)

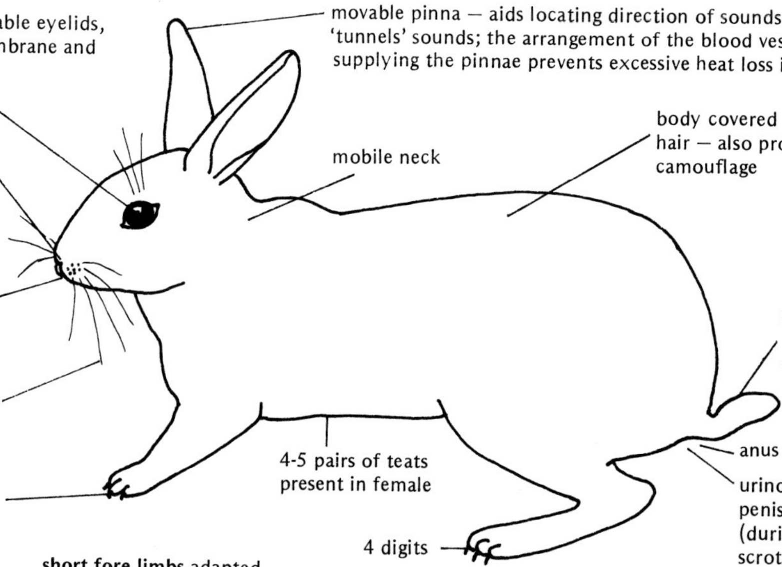

Notes

1. Found in large numbers occupying complex ramifying communal burrows (warrens) in woodlands, grassland, sand dunes, salt marshes, mountains, moorlands and cliffs.
2. Feeds mainly at night, remaining in burrows during the day.
3. Hard dry faecal pellets are deposited at specific sites some distance from the burrow. In the burrow the rabbit indulges in coprophagy; it eats its soft moist faeces. Therefore food passes through the alimentary canal twice and the maximum possible nutriment is obtained from it.
4. Preyed on by foxes, badgers, cats, weasels, dogs and man. Relies on acute senses of hearing and smell to detect danger. Long hind limbs enable fast running and leaping to aid escape. Eyes lateral and prominent, giving 360° field of view.
5. Extraordinary fertility to compensate for the high predation; 3–4 litters a year, 4–10 young per litter, sexual maturity reached at 5 months. (A single pair of rabbits could produce 13 million descendants in 3 years). The young are born in separate nesting burrows.

The main orders of Eutherian Mammals are:

INSECTIVORA

Shrew (x ¾)

General characteristics

Usually small with long tapering snout and abundant vibrissae.

Feed on insects.

Usually nocturnal.

Numerous teeth; incisors often act like forceps, cheek teeth sharply pointed and interlocking. (Some dagger-like, some saw-like, others blade-like.)

Voracious appetite to compensate for heat loss due to small size.

Plantigrade, with 5 clawed digits on each limb.

e.g. hedgehogs, moles, shrews

CHIROPTERA

Long eared bat (x ⅓)

General characteristics

Adapted for flying.

Usually nocturnal.

Feed on insects or fruit or are blood suckers. Insects are caught with the mouth while the bat is in flight; they may be stored in the pocket-like tail end of the wing.

Teeth small; molars with sharp cusps in insect eaters, flattened in fruit eaters.

Forelimbs modified to support the wings – digits elongated. Wings composed of skin.

Enormous body surface results in heat loss and water loss; feed intensely and drink frequently.

Hind feet prehensile. Hang from hind feet when at rest. Can climb and crawl, but cannot walk well.

Emit ultrasonic sounds, and avoid obstacles by echo-location.

Single offspring, carried by mother.

Sleep in clusters to conserve heat. Hibernate in winter.

e.g. bats

PRIMATES

Chimpanzee (x 1/12)

General characteristics

Primitively arboreal.

Omnivorous or herbivorous.

Teeth unspecialised.

Plantigrade, with tendency to bipedal gait.

Tail, if present, may be prehensile.

Manipulative sensitive hands with opposable thumbs. Flat nails, not claws.

Eyes face forwards, stereoscopic colour vision (sight the dominant sense). Sense of smell reduced.

Pinnae reduced.

Great development of cerebral hemispheres. Highly intelligent, capable of reasoned thought.

Mating occurs throughout the year.

Tend to live in family groups, tribes, etc.

e.g. lemurs, monkeys, apes, humans

LAGOMORPHA

Hare (x $^1/_{10}$)

General characteristics

Moderately sized gnawing herbivores with long ears and short tail.

Mainly nocturnal.

Herbivorous, with chisel-like incisors (2 pairs upper incisors, one pair behind the other), diastema, grinding cheek teeth. Jaw action backwards/forwards. All teeth open rooted.

Plantigrade. Long hind limbs enable fast running and leaping, fore limbs used for burrowing (in rabbits only).

Large caecum. Coprophagous; faeces eaten, food passes through gut twice.

Several offspring produced several times a year.

e.g. rabbits, hares

RODENTIA

Woodmouse (x ¾)

General characteristics

Usually small sized gnawing mammals.

Primarily herbivorous, but some omnivorous. Chisel-like incisors with enamel only on front surface, open rooted. Diastema, grinding cheek teeth. Jaw action backwards/forwards.

Plantigrade. Fore limbs often used for handling food when eating.

Cheek pouches often present. Food can be gathered quickly and eaten later in safety.

Many hoard stores of food. Many burrow, many build.

Large caecum.

Breed and multiply rapidly; unrivalled by any other group except rabbits.

Very widely distributed in diverse habitats.

e.g. rats, guinea-pigs, mice, voles, porcupines, squirrels, beavers, hamsters

CETACEA

Dolphin (x $^1/_{30}$)

General characteristics

Aquatic, marine.

Carnivorous, feeding either on plankton, or fish or cephalopods.

Plankton-feeders possess a filtering device (baleen) of modified hairs, replacing teeth. In fish-eaters the teeth are conical and all alike.

No sense of smell.

Streamlined shape; no pinnae, no neck (cervical vertebrae compressed), no hind limbs, no hair.

Blubber (fat) stored in dermis, insulating body.

Fore limbs modified to form flippers (any other fins are merely folds of skin).

Complex stomach with 'gizzard'.

Tail with 2 horizontal flukes, provides the propulsive force in swimming.

Respiratory and vascular system highly modified for life in water, especially for diving; enormous lungs, no connection between mouth and lungs (allowing feeding underwater), nostrils replaced by dorsal blowhole(s).

e.g. whales, dolphins, porpoises

CARNIVORA

Cat (x $\frac{1}{6}$)

General characteristics

Usually medium/large-sized carnivorous hunters.

Strong agile lean body.

Long strong canine teeth forming fangs, 1 pair carnassial cheek teeth present for shearing flesh.

Only vertical movements of the jaw are possible, enabling a powerful bite but no grinding or chewing of food.

Usually digitigrade, allowing stealthy movement. Sharp strong claws present, may be retractile.

Keen senses of smell and hearing. Sight may also be well developed. Forward facing eyes.

Intestine short, no caecum.

Cunning and courageous behaviour.

Being large animals with a high protein diet allows leisure for play.

e.g. dogs, cats and relatives, otters, badgers, weasels, foxes, wolves, bears, hyenas

SUB-ORDER PINNIPEDIA

Seal (x $\frac{1}{25}$)

General characteristics

Marine carnivores, feeding on fish, crustaceans and molluscs.

Streamlined body, but distinct neck present.

Blubber present, but body also covered by short fur.

Upper canines may be fang-like. No carnassial teeth.

Fore and hind limbs modified to form flippers, but digits and claws remain. Fore limbs enable movement on land.

Tail vestigial; hind limbs provide propulsive force in swimming.

Hearing acute.

Birth usually occurs on land. Single offspring.

e.g. seals, walruses

PROBOSCIDEA

African elephant (x $\frac{1}{40}$)

General characteristics

Enormous herbivores of forest and savannah.

Nose and upper lip elongated to form a very strong, very sensitive prehensile trunk, used for collecting food and for many other varied purposes.

Upper incisors form tusks. No lower incisors, no canines, premolars soon shed. Enormous grinding molars, developed in series and used one after the other.

Thick almost hairless skin. Enormous pinnae used for dissipating body heat, discouraging flies and for threatening.

Massive columnar limbs.

Plantigrade, with 5 digits on each limb.

Eat enormous amounts of food and must constantly move to find new pastures.

Live in herds. Show considerable mutual concern.

Grow throughout life.

Usually single offspring.

e.g. Elephants

PERISSODACTYLA

Horse (x $^1/_{30}$)

General characteristics

Large hoofed herbivores, with only 1 or 3 digits per limb touching the ground (i.e. odd-toed ungulates). Adapted for fast running on open ground.

Feed by grazing and cropping. Large sharp incisors for cutting grass, large premolars and molars providing a considerable grinding surface with ridges transverse to the jaw bone. Jaw action backwards/forwards. Diastema present.

Limbs lengthened by the elongation of the metacarpals and metatarsals, raising the animal up on its toes, and thereby attaining the maximum speed in running. i.e. unguligrade (walking on the tips of the digits which are reduced in number and provided with hooves).

Powerful hind quarters. Limb action restricted to forwards/backwards.

Long neck to reach food and afford a good look out.

Long face to accommodate the large cheek teeth and diastema. Large pinnae to detect direction of sounds.

Large caecum (no rumen).

Usually unaggressive.

e.g. horses, zebras, donkeys, rhinoceroses

ARTIODACTYLA

Fallow deer (x $^1/_{20}$)

General characteristics

Large hoofed usually ruminant herbivores adapted for fast running on open ground. Similar to the perissodactyls in many characters, most of which have been evolved independently (i.e. parallel evolution).

Usually only the 3rd and 4th digits are present and are hoofed (cloven hooves), i.e. even-toed ungulates.

Incisors often absent, being replaced by a horny cropping pad. Canines may form tusks. Premolars smaller than molars, ridges of grinding surface are parallel to the jaw bone and the jaw action is sideways.

The tongue is long, mobile and protrusible, often with horny papillae.

Unguligrade.

Horns common.

Rumen usually present; animals 'chew the cud'.

e.g. ruminants: cattle, sheep, deer, giraffes, goats, (also camels, which are not true ruminants)
non-ruminants: pigs, hippopotamuses

Glossary

See page numbers in italic for examples.

aboral, furthest away from the mouth

acellular, the term applied to unicellular organisms which are complex in structure and function, *9*

acoelomate, lacking a coelom

actin, protein filament concerned with contractile mechanisms in cells

aerobic, requiring oxygen from the environment

altricial bird, having young which are naked and helpless after hatching, *80*

ambulacral, concerned with walking

amitosis, division of the nucleus by simple constriction into two halves, *14*

anaerobic, not requiring the presence of oxygen in the surroundings

anatomy, the internal structure of organisms

analogous, of similar function but not of similar origin

animalcule, microscopic animal

antibiotic, substance secreted by a micro-organism into its surroundings where it is toxic to other species, especially bacteria

antibody, specific protein produced by an animal cell in response to the presence of an antigen (foreign substance)

anus, terminal opening of the gut

apodous, lacking legs

artificial classification, the grouping of organisms according to criteria which bear no relation to phylogeny, i.e. with reference to analogous rather than homologous structures

atrium, i) chamber surrounding pharynx in protochordata
ii) alternative name for auricle of the heart

auricle, chamber of the heart which receives blood from veins

autotrophic, independent of any organic materials from the environment

bacteriophage, virus which infects bacteria

basal granule/body, structure similar to centriole, to which a cilium or flagellum is anchored

bilateral symmetry, symmetry about a single plane, so that there is one way only in which the body can be divided into two similar halves

binominal nomenclature, universally accepted system of naming species by two latin words, the first denoting its genus and the second its species

blood sinus, i) enormously expanded vein
ii) division of haemocoel in arthropods

branchial cleft, visceral cleft modified to house gills, *67*

buccal cavity, mouth cavity, lined by ectoderm, *40, 69*

byssus, threads secreted by some bivalves for adherence to rocks

caecum, a blind branch of the gut

carapace, protective shield of exoskeleton covering the dorsal surface of the cephalothorax in some crustaceans

carnassial teeth, a pair of strong blade-like shearing teeth which act like scissors. Found in Carnivora. Usually the last upper premolar and the first lower molar

cephalisation, the specialisation of the anterior end of the body to form a head

chaeta, chitinous bristle, *43*

cilium, short mobile thread-like extension of the cytoplasm of a cell, occurring in enormous numbers and beating rhythmically, *13, 17, 31, 38*

cloaca, common genital and intestinal opening

closed blood system, circulatory system where the blood is retained within a system of continuous vessels

coelomate, possessing a coelom, a body cavity originating by the splitting of the mesoderm into inner and outer layers in triploblast animals

coenocytic, multinucleate mass of cytoplasm, *6*

colonial, organised association of incompletely separate interdependent individuals, *20, 23, 26*

coprophagy, the eating of faeces

compound eye, eye composed of numerous separate image-forming elements. Found only in insects and crustaceans

conjugation, a type of sexual reproduction where the gametes consist of the entire contents of unspecialised cells and cannot be distinguished by appearance as male or female, *9, 14*

cryptic colouration, camouflage

ctenidium, comb-like gill of molluscs

cuticle, external non-cellular layer secreted by the epidermis, *32, 54*

diastema, toothless gap between the incisors and premolars enabling the leaves which are being cropped to be kept separate from those being masticated, *86, 88*

digitigrade, walking on the ventral surface of the digits (i.e. on the toes), *87*

dioecious, having separate male and female individuals, *22, 25*

diploblastic, having the body organised from only two embryonic cell layers (ectoderm and endoderm), *22*

diploid, possessing two sets of chromosomes in the cell nucleus i.e. the chromosomes occur in homologous pairs

direct development, development without a larval form, *80*

distal, situated away from the point of origin or attachment

diurnal, active by day

double circulation, blood system which is divided into a pulmonary and a systemic circulation, and the heart is divided (or partly divided) into two separate pumps

ecdysis, the shedding of the outer layer of cuticle in arthropods (moulting)

ectoderm, superficial layer of cells in animal embryo. Gives rise to epidermis, nervous tissue and nephridial system

ectoplasm, outer clear layer of cytoplasm of a cell – semi-solid

ectothermic, 'cold-blooded' – applied to animals whose body temperature can be raised only from external sources, *72, 75, 78*

elytron, hardened fore-wing of insect, especially beetles, protecting the delicate hind-wing beneath it

embryo, developing organism derived from zygote

endoderm, layer of embryonic cells which after invagination gives rise to the gut and associated organs

endoplasm, internal granulous cytoplasm of a cell, possessing organelles. Usually more fluid than ectoplasm

endostyle, ciliated groove in ventral wall of pharynx of protochordates, concerned in ciliary feeding

endopterygote insect, insect with complete metamorphosis, i.e. the juvenile form is a wingless larva unlike adult in both appearance and mode of life. The abrupt change into the adult form takes place during a pupal stage

endothermic, 'warm-blooded' – applied to animals which can generate sufficient heat within their bodies to maintain a constant high body temperature regardless of external conditions, *80, 83*

enzyme, specific organic (protein) catalyst produced by a living cell, *6*

epiphysis, separately ossified end of a mammalian bone, parted from the rest of the bone by a layer of cartilage until growth is complete

eucaryotic, cell where the genetic material is incorporated in chromosomes within a nucleus and where membrane-bounded organelles are present, *6, 9*

exopterygote insect, insect with incomplete metamorphosis, i.e. the juvenile form is a nymph whose wings develop gradually by small increments at each moult

evolution, the process of very slow and gradual change which occurs in the descendent generations of populations of organisms by a process of Natural Selection

fission, asexual reproduction by the division of the body into two or more equal parts, *6, 9, 14*

flagellum, long mobile thread-like extension of the cytoplasm of a cell, with lashing or undulating beat. Usually only one or few per cell, *6, 11, 21*

gamete, haploid reproductive cell whose fusion with another gamete produces a diploid zygote which develops into a new individual, *15*

ganglion, small mass of nervous tissue composed principally of cell bodies of neurons, *30, 45*

haemocoel, body cavity containing blood, derived from blastocoel of embryo (not coelom), *45*

haltere, modified hind wing of diptera, concerned with maintaining balance during flight

haploid, possessing one set of (unpaired) chromosomes in the cell nucleus, *14*

herbivore, plant eater

hermaphrodite, bisexual, a single individual producing both male and female gametes, *30, 41, 63*

heterocercal tail, caudal fin with ventral lobe more extensive and flexible than upper lobe, *72, 73*

heterodont, having teeth of varying shapes with different functions, *83*

heterotrophic, requiring organic material from the environment

holophytic, feeding like a typical plant, i.e. involving photosynthesis

holozoic, feeding like a typical animal, i.e. ingesting solid organic food which is subsequently digested

homodont, dentition where all the teeth are similar, *78*

homoiothermic, 'warm-blooded' – see **endothermic**

homologous, of similar origin

hyostylic jaw suspension, upper and lower jaws attached indirectly to the cranium, by way of the hyomandibula of the hyoid arch, *72*

hypostome, oral cone of hydroid coelenterate, the tentacles arising from around the base of the cone and the mouth at the apex

imago, fully grown sexually mature insect

ingestion, the taking of food from the environment to within the confines of the body

instar, stage between successive ecdyses during the larval development of an insect

intercellular, between cells

internal nares, paired openings from the vertebrate olfactory organs into the mouth

intracellular, within a cell

labellum, proboscis of a fly

larva, juvenile form differing from adult in both appearance and mode of life, and leading an independent existence. Undergoes metamorphosis to become adult, *34, 54*

lateral line, system of sense organs responsive to pressure changes in water. Situated in canals just below the skin, arranged in a complicated pattern in the head and in a line down each side of the body of aquatic vertebrates

lateral plate mesoderm, unsegmented mesoderm on the lateral and ventral sides of a vertebrate embryo within which the coelom arises. Gives rise to smooth muscle, cardiac muscle and coelomic epithelium, and to striped muscle and skeleton of tetrapod limb

lysis, rupture of plasma membrane, destroying cell, with loss of cell contents, *5*

madroporite, (sieve plate), opening of water vascular system of echinoderms

Malpighian tubule, excretory gland opening into the hind gut of terrestrial arthropods

manubrium, the projection from the sub-umbrella surface of a medusoid coelenterate, with the mouth at the apex

meiosis, (reduction division), type of nuclear division occurring when a diploid cell divides to produce four haploid daughter cells. Occurs in the life cycle of all sexually reproducing organisms, *14*

mesoderm, layer of cells of triploblast embryo which is invaginated and from which is derived muscle tissue, blood, connective tissue, etc.

mesogloea, non-cellular jelly-like material between the endoderm and ectoderm of a coelenterate

metabolism, all the chemical processes occurring within an organism

metameric segmentation, organisation of the body into a series of similar interdependent segments of the same age and of constant number, *40*

metamorphosis, the series of changes which take place as a larva transforms into the adult form, *53, 54*

microvillus, minute finger-like projection from the surface of a cell

mitosis, type of nuclear division occurring when a cell divides during growth and asexual reproduction. Each daughter cell is genetically identical to the parent cell, *14, 16*

monoecious, hermaphrodite

mycelium, mass of hyphae of a fungus, excluding the reproductive bodies

myosin, protein filament concerned with contractile mechanisms in cells

natural classification, system of grouping organisms so that those with the maximum possible number of attributes in common are placed together

nephridium, ectodermal excretory and osmo-regulatory organ of many invertebrates

nictitating membrane, transparent third eyelid attached at the inner corner of the eye and moving horizontally across the eye. Well developed in reptiles and birds, and common in other tetrapods

nocturnal, active by night

notochord, axial skeletal rod present at some stage in the development of chordates. Replaced by the vertebral column in vertebrates

nymph, juvenile stage of exopterygote insect, resembling adult in many ways, but lacking wings

ocellus, simple eye. Photoreceptor with no image formation, *50*

ommatidium, single element of compound eye

open blood system, circulatory system where the blood is not confined to vessels, the body cavity being a haemocoel, *45, 62*

open-rooted tooth, growing continuously throughout the life of the animal, *86*

operculum, lid or cover, especially of gills

osmo-regulation, the control of the water balance in the body

parapodium, muscular locomotory projection from the side of the body of a polychaete worm, occurring in segmental pairs

parasite, organism living in or on another living organism (host) from which it derives its food, *4, 7, 15, 34*

parthenogenesis, the development of an unfertilized egg (oocyte) into a new individual, *47, 60*

pathogenic, causing disease

pellicle, living firm elastic 'skin' of protozoon, internal to the plasma membrane, conferring definite shape

peristalsis, successive waves of contraction passing along the muscular walls of a tubular organ

pellagic, inhabiting the open waters of sea or lakes, as opposed to the bottom. Includes surface-floating plankton and swimming nekton

phagocytosis, the engulfing of particles of food by a cell by means of pseudopodia

phototaxis, a directional locomotory movement of a micro-organism in response to the stimulus of light

phylogeny, the evolutionary history of groups of organisms

physiology, the study of the processes which occur in living organisms

planktonic, small and microscopic organisms that drift freely in surface waters

plantigrade, walking on the ventral surface of the whole foot, *86*

plasmagel, ectoplasm

plasmasol, endoplasm

pleopod, abdominal appendage of an arthropod

poikilothermic, 'cold-blooded', see **ectothermic**

polymorphism, the existence of morphologically different types of adult individuals in a single species, *22*

polypodous, possessing many legs

prehensile, able to grasp

primitive, originating at a relatively early stage of evolutionary history

proboscis, i) insect mouth parts specialised for sucking up fluid
ii) trunk of elephant

procaryotic, primitive type of cell with no defined nucleus or other membrane-bounded organelles, as in bacteria

proleg, short, fleshy, unjointed walking appendage borne on the abdomen of a caterpillar

prostomium, presegmental region at the anterior end of an annelid worm, anterior to the mouth

Protista, group (Kingdom) which includes all organisms of simple biological organisation

pseudopodium, temporary protrusion of the cytoplasm of an amoeboid cell, concerned with locomotion and phagocytosis

pupa, protected stage between the larva and imago of an endopterygote insect during which metamorphosis occurs

putrefaction, the decomposition of protein-containing material

pygidium, terminal unsegmented region of an annelid worm, in which the anus is situated

pyrenoids, spherical bodies embedded in chloroplasts of green algae, concerned with starch synthesis and storage

radially symmetrical, having similar parts arranged around a central axis

radula, horny toothed strip in the mouth of molluscs, used for rasping food

saprophytic, feeding on non-living organic matter, which is digested externally and absorbed in solution, *7, 12*

schizogony, rapid asexual multiplication by multiple fission, common in parasitic protozoa

sclerite, plate of hardened chitin in the cuticle of arthropods. Flexible regions between adjacent sclerites form the joints of the skeleton

segmented, see **metameric segmentation**

seminal vesicle, organ for storage of sperm

sessile, i) permanently fixed to the substratum
ii) without a stalk (of compound eye)

sexual reproduction, the production of new individuals by a process involving the union of two gametes

simple eye, see **ocellus**

single circulation, blood system of fish and primitive chordates where the heart pumps only deoxygenated blood

somite, one of the segmental blocks of mesoderm flanking the notochord in the vertebrate embryo

specialised, having changed significantly during the course of evolution and become highly adapted for a particular mode of life or habitat

spiracle, i) reduced first gill slit in elasmobranch fish, through which water can enter during breathing
ii) external opening of the tracheal system in insects
iii) the opening from the opercular chamber of a tadpole

spiral valve, spiral fold in the lining of the intestine of elasmobranch fish, increasing the absorptive surface

statocyst, organ of balance consisting of a sac lined by sensory cells and containing granules of lime or sand, *26, 49*

sub-littoral, below the shore line

swim bladder, gas-filled hydrostatic organ in bony fish which changes the specific gravity of the body according to depth

symbiosis, a very close association between two different species where both derive benefit from the association, *6*

synapse, the gap across which impulses are relayed at the point of junction between successive neurons, *24*

systematics, the study of biological classification with reference to the course of evolution within the various groups, and any or all the relationships among them

systemic circulation, general body circulation

taxonomy, the study of the rules, principles and practice of classifying organisms

telson, terminal abdominal segment of an arthropod

tetrapod, four-limbed vertebrate

triploblastic, having the body organised from three embryonic layers of cells; ectoderm, mesoderm and endoderm, *19, 29*

tube-feet, hollow extensile appendages connected to the water vascular system of echinoderms

unguligrade, walking on the tips of the digits, which are reduced in number and provided with immensely enlarged nails (hooves), *88*

vas deferens, tube conveying sperm from the testis to the exterior

vector, secondary host of a parasite which transmits the parasite directly from one primary host to another, *34*

vestigial, reduced in the course of evolution to small remains, and with no function

zygote, diploid cell formed by the fusion of two gametes, *16*